The Rapid Evaluation of Potential Fields in Particle Systems

ACM Distinguished Dissertations

The Rapid Evaluation of Potential Fields in Particle Systems

Leslie Greengard

The MIT Press
Cambridge, Massachusetts
London, England

Publisher's Note

This format is intended to reduce the cost of publishing certain works in book form and to shorten the gap between editorial preparation and final publication. Detailed editing and composition have been avoided by photographing the text of this book directly from the author's prepared copy.

Printed and bound in the United States of America

Library of Congress Cataloging-in-Publication Data

Greengard, Leslie.
  The rapid evaluation of potential fields in
particle systems.

  (ACM distinguished dissertations ; 1987)
  Thesis—
  Bibliography: p.
  Includes index.
  1. Potential, Theory of. 2. Particles. 3. Algo-
rithms. 4. Mathematical physics. I. Title.
II. Series.
QC20.7.P67G74   1988   530.1'557   88-600
  ISBN 0-262-07110-X
  ISBN: 0-262-57192-7 (Paperback)

Contents

List of Figures

List of Tables

Preface

The evaluation of Coulombic or gravitational interactions in large-scale ensembles of particles is an integral part of the numerical simulation of a large number of physical processes. Examples include celestial mechanics, plasma physics, the vortex method in fluid dynamics, and molecular dynamics. In a typical application, a numerical model follows the trajectories of a number of particles moving in accordance with Newton's second law of motion in a field generated by the whole ensemble. In many situations, in order to be of physical interest, the simulation has to involve thousands of particles (or more), and the fields have to be evaluated for a large number of configurations. Unfortunately, an amount of work of the order $O(N^2)$ has traditionally been required to evaluate all pairwise interactions in a system of N particles, unless some approximation or truncation method is used. As a result, large-scale simulations have been extremely expensive in some cases, and prohibitive in others.

This thesis presents an algorithm for the rapid evaluation of the potential and force fields in large-scale systems of particles. In order to evaluate all pairwise Coulombic interactions of N particles to within round-off error, the algorithm requires an amount of work proportional to N, and this estimate does not depend on the statistics of the distribution. In practice, speedups of three to four orders of magnitude may be expected in a system of a million particles, rendering previously prohibitive simulations feasible.

Both two and three dimensional versions of the algorithm have been constructed, and we will discuss their applications to several problems in physics, chemistry, biology, and numerical complex analysis.

Acknowledgements

I would like to express my gratitude to my advisors, Drs. Martin Schultz and Vladimir Rokhlin. Dr. Schultz has been very supportive during my time as a graduate student. His encouragement and fund of knowledge have been a great benefit. Much of the work contained in this thesis is based on a point of view of applied mathematics and numerical analysis which I have acquired from Dr. Rokhlin. I am enormously indebted to him for this.

I would like to thank Dr. Bill Gropp of the Computer Science Department at Yale for many useful discussions, and Dr. Charles Peskin of the Courant Institute for being a reader and for his continued interest.

Finally, I thank Jean Carrier for his help in developing the adaptive algorithm described in section 2.5 .

The Rapid Evaluation of Potential Fields in Particle Systems

1 Introduction

The study of physical systems by particle simulation is well-established in a number of fields. It is becoming increasingly important in others. A classical example is celestial mechanics, but much recent work has been done in formulating and studying particle models in plasma physics, fluid dynamics, and molecular dynamics [24].

There are two major classes of simulation methods. Dynamical simulations follow the trajectories of N particles over some time interval of interest. Given initial positions $\{x_i\}$ and velocities, the trajectory of each particle is governed by Newton's second law of motion:

$$m_i \, \frac{d^2 x_i}{dt^2} = -\nabla_i \Phi \qquad \text{for } i = 1, ..., N \ ,$$

where m_i is the mass of ith particle, and the force is obtained from the gradient of a potential function Φ. When one is interested in an equilibrium configuration of a set of particles rather than their time-dependent properties, an alternative approach is the Monte Carlo method. In this case, the potential function Φ has to be evaluated for a large number of configurations in an attempt to accurately describe the potential surface.

In a typical application, the potential has the form

$$\Phi = \Phi_{near} + \Phi_{external} + \Phi_{far} \ ,$$

where Φ_{near} is a rapidly decaying function of distance (e.g. the Van der Waals potential in chemical physics), $\Phi_{external}$ is independent of the number of particles (e.g. an applied external electrostatic field), and Φ_{far}, the far-field potential, is Coulombic or gravitational. Such models describe classical celestial mechanics and many problems in plasma physics and molecular dynamics. In the vortex method for incompressible fluid flow calculations [12], an important and expensive portion of the computation has the same formal structure (the stream function and the vorticity are related by Poisson's equation).

In a system of N particles, the calculation of Φ_{near} requires an amount of work proportional to N, as does the calculation of $\Phi_{external}$. The decay of the Coulombic or gravitational potential, however, is sufficiently slow that all interactions must be accounted for, resulting in CPU time requirements of the order $O(N^2)$. In this dissertation, a method is presented for the rapid (order $O(N)$) evaluation of these interactions for all particles to within round-off error.

1.1   Brief History

There have been a number of previous efforts aimed at reducing the computational complexity of the N-body problem. Particle-in-cell methods [24] have received careful study and are used with much success, most notably in plasma physics. Assuming the potential satisfies Poisson's equation, a regular mesh is layed out over the computational domain and the method proceeds by:

1. interpolating the source density at mesh points,

2. using a "fast Poisson solver" to obtain potential values on the mesh,

3. computing the force from the potential and interpolating to the particle positions.

The complexity of these methods is of the order $O(N + M \, logM)$, where M is the number of mesh points. The number of mesh points is usually chosen to be proportional to the number of particles, but with a small constant of proportionality so that $M \ll N$. Therefore, although the asymptotic complexity for the method is $O(N \, logN)$, the computational cost in practical calculations is usually observed to be proportional to N. Unfortunately, the mesh provides limited resolution, and highly non-uniform source distributions cause a significant degradation of performance. Further errors are introduced in step (3) by the necessity for numerical differentiation to obtain the force.

To improve the accuracy of particle-in-cell calculations, short-range interactions can be handled by direct computation, while far-field interactions are obtained from the mesh, giving rise to so-called particle-particle/particle-mesh (P^3M) methods [24]. For an implementation of these ideas in the context of vortex calculations, see [5]. While these algorithms still depend for their efficient performance on a reasonably uniform distribution of particles, in theory they do permit arbitrarily high accuracy to be obtained. As a rule, when the required precision is relatively low, and the particles are distributed more or less uniformly in a rectangular region, P^3M methods perform satisfactorily. However, when the required precision is high (as, for example, in the modeling of highly correlated systems), the CPU time requirements of such algorithms tend to become excessive.

Appel [7] introduced a "gridless" method for many-body simulation with a computational complexity estimated to be of the order $O(N \, logN)$. It relies on using a monopole (center-of-mass) approximation for computing forces over large distances and sophisticated data structures to keep track of which particles are sufficiently clustered to make the approximation valid. For certain types of problems, the method achieves a dramatic speed-up compared to the naive $O(N^2)$ approach. It is less efficient when the distribution of particles is relatively uniform and the required precision is high.

1.2   Outline of the Dissertation

The algorithms presented here make use of multipole expansions to compute potentials or forces to whatever accuracy is required. Portions of the work described below have been published previously [20,21,11]. The approach taken is similar to the one introduced in [36] for the solution of boundary value problems for the Laplace equation.

In chapter 2, we consider potential problems in two dimensions and begin with the introduction of the necessary mathematical preliminaries. A fast multipole algorithm is then developed for the evaluation of the potentials and forces in large-scale systems of particles randomly distributed in a square domain. This method requires an amount of work proportional to N to evaluate all pairwise interactions in a system of N charges. The chapter ends with a description of an adaptive version of the algorithm whose CPU time requirements are proportional to N and independent of the statistics of the charge distribution.

In Chapter 3, three-dimensional systems of particles are considered. The mathematical foundation of the method in this case is the theory of spherical harmonics, which is developed in some detail. In particular, two generalizations of the classical addition theorem for Legendre polynomials (Theorems 3.5.1 and 3.5.2) are formulated and proved. They appear to have been previously unknown, and are needed for the development of efficient translation operators which are critical features of the algorithm. It should be noted, however, that despite the increased mathematical complexity of the three-dimensional case, the framework of the fast multipole algorithm is the same as in two dimensions.

In chapter 4, we present numerical results demonstrating the actual performance of the method, and in chapter 5, we briefly outline some applications and generalizations.

2 Potential Fields in Two Dimensions

Many physical processes are adequately described by two-dimensional models, and this fact is widely exploited in computer simulations. From the computational point of view, reduction of the dimensionality of the problem has two major advantages: fewer particles are normally required to obtain a physically meaningful model of a two-dimensional process than of its three-dimensional counterpart, and numerical methods for calculations in two dimensions are better developed and easier to implement. Moreover, the display and interpretation of three-dimensional results pose problems almost non-existent in two dimensions. Certain processes in the physical world, however, simply can not be approximated by two-dimensional models. In such cases, full three-dimensional simulations have to be performed, with the help of appropriate numerical tools. We postpone the consideration of such problems to the next chapter. For the moment, we assume that the potential and force fields are known to be independent of one of the Cartesian coordinates, say the third coordinate z. In these situations, the governing equation for potential problems is the two-dimensional Laplace equation

$$\nabla^2 \Phi = \frac{\partial^2 \Phi}{\partial x^2} + \frac{\partial^2 \Phi}{\partial y^2} = 0 \; . \tag{2.1}$$

Functions which satisfy (2.1) are referred to as harmonic functions.

The physical model we will consider in this chapter consists of a set of N charged particles, lying in the (x, y)-plane. In such two dimensional systems, the force of attraction between two particles varies as the inverse first power of the distance between them. More specifically, if a two-dimensional point charge is located at the point $(x_0, y_0) = \mathbf{x}_0 \in \mathrm{R}^2$, then for any $\mathbf{x} = (x, y) \in \mathrm{R}^2$ with $\mathbf{x} \neq \mathbf{x}_0$, the potential and electrostatic field due to this charge are described by the expressions

$$\phi_{\mathbf{x}_0}(x, y) = -\log(\| \mathbf{x} - \mathbf{x}_0 \|) \tag{2.2}$$

and

$$E_{\mathbf{x}_0}(x, y) = \frac{(\mathbf{x} - \mathbf{x}_0)}{\| \mathbf{x} - \mathbf{x}_0 \|^2} \tag{2.3}$$

respectively. In the remainder of this chapter, all particles are assumed to be such two-dimensional ones. Section 2.1 below develops a series expansion of the field due to an arbitrary distribution of charge, while section 2.2 describes certain translation operators which will allow us to manipulate both far field and local expansions in the manner required by the fast algorithm.

2.1   The Field of a Charge

It is well-known that the function $\phi_{\mathbf{x}_0}$, defined above, is harmonic in any region not containing the point $\mathbf{x}_0$. Moreover, for every harmonic function u, there exists an analytic function $w : \mathbb{C} \to \mathbb{C}$ such that $u(x, y) = Re(w(x, y))$, and w is unique except for an additive constant. In this chapter, we will work with analytic functions, making no distinction between a point $(x, y) \in \mathbb{R}^2$ and a point $x + iy = z \in \mathbb{C}$. We note that

$$\phi_{\mathbf{x}_0}(\mathbf{x}) = Re(-\log(z - z_0)), \tag{2.4}$$

and, following standard practice, we will refer to the analytic function $\log(z)$ as the potential due to a charge. As we develop expressions for the potential due to more complicated charge distributions, we will continue to use complex notation, and will refer to the corresponding analytic functions themselves as the potentials. The following lemma is an immediate consequence of the Cauchy-Riemann equations.

LEMMA 2.1.1 *If $u(x, y) = Re(w(x, y))$ describes the potential field at (x, y), then the corresponding force field is given by*

$$\nabla u = (u_x, u_y) = (Re(w'), -Im(w')), \tag{2.5}$$

where w' is the derivative of w.

The following lemma is used in obtaining the multipole expansion for the field due to m charges.

LEMMA 2.1.2 *Let a point charge of intensity q be located at z_0. Then for any z such that $|z| > |z_0|$,*

$$\phi_{z_0}(z) = q \log(z - z_0) = q \left(\log(z) - \sum_{k=1}^{\infty} \frac{1}{k} \left(\frac{z_0}{z} \right)^k \right). \tag{2.6}$$

Proof: Note first that $\log(z - z_0) - \log(z) = \log\left(1 - \frac{z_0}{z}\right)$ and that $\left|\frac{z_0}{z}\right| < 1$. The lemma now follows from the expansion

$$\log(1 - w) = (-1) \sum_{k=1}^{\infty} \frac{w^k}{k} \quad , \tag{2.7}$$

which is valid for any w such that $|w| < 1$.

THEOREM 2.1.1 (**Multipole Expansion**) *Suppose that m charges of strengths $\{q_i, \ i = 1, ..., m\}$ are located at points $\{z_i, \ i = 1, ..., m\}$, with $|z_i| < r$. Then for any $z \in \mathbb{C}$ with $|z| > r$, the potential $\phi(z)$ induced by the charges is given by*

$$\phi(z) = Q \log(z) + \sum_{k=1}^{\infty} \frac{a_k}{z^k}, \tag{2.8}$$

where

$$Q = \sum_{i=1}^{m} q_i \quad and \quad a_k = \sum_{i=1}^{m} \frac{-q_i z_i^k}{k}. \tag{2.9}$$

Furthermore, for any $p \geq 1$,

$$\left| \phi(z) - Q \log(z) - \sum_{k=1}^{p} \frac{a_k}{z^k} \right| \leq \alpha \left| \frac{r}{z} \right|^{p+1} \leq \left(\frac{A}{c-1} \right) \left(\frac{1}{c} \right)^p \tag{2.10}$$

where

$$c = \left| \frac{z}{r} \right| \qquad A = \sum_{i=1}^{m} |q_i| \quad , \ and \quad \alpha = \frac{A}{1 - |\frac{r}{z}|}. \tag{2.11}$$

Proof: The form of the multipole expansion (2.8) is an immediate consequence of the preceding lemma and the fact that $\phi(z) = \sum_{i=1}^{m} \phi_{z_i}(z)$. To obtain the error bound (2.10), observe that

$$\left| \phi(z) - Q \log(z) - \sum_{k=1}^{p} \frac{a_k}{z^k} \right| = \left| \sum_{k=p+1}^{\infty} \frac{a_k}{z^k} \right| \tag{2.12}$$

Substituting for a_k the expression in (2.9), we have

$$\left| \sum_{k=p+1}^{\infty} \frac{a_k}{z^k} \right| \leq A \sum_{k=p+1}^{\infty} \frac{r^k}{k |z|^k} \leq A \sum_{k=p+1}^{\infty} \left| \frac{r}{z} \right|^k = \alpha \left| \frac{r}{z} \right|^{p+1} = \left(\frac{A}{c-1} \right) \left(\frac{1}{c} \right)^p \tag{2.13}$$

In particular, if $c \geq 2$, then

$$\left| \phi(z) - Q \log(z) - \sum_{k=1}^{p} \frac{a_k}{z^k} \right| \leq A \left(\frac{1}{2} \right)^p \tag{2.14}$$

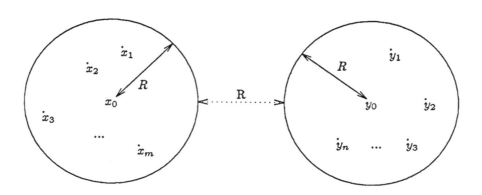

Figure 2.1
Well-separated sets in the plane.

Finally, we demonstrate with a simple example how multipole expansions can be used to speed up calculations with potential fields. Suppose that charges of strengths $q_1, q_2, ..., q_m$ are located at the points $x_1, x_2, ..., x_m \in \mathbb{C}$ and that $\{y_1, y_2, ..., y_n\}$ is another collection of points in $\mathbb{C}$ (Figure 2.1).

We say that the sets $\{x_i\}$ and $\{y_i\}$ are *well-separated* if there exist points $x_0, y_0 \in \mathbb{C}$ and a real $r > 0$ such that

$$
\begin{aligned}
|x_i - x_0| &< r & \text{for all } i = 1, ..., m\,, \\
|y_j - y_0| &< r & \text{for all } j = 1, ..., n\,, \quad \text{and} \\
|x_0 - y_0| &> 3r.
\end{aligned}
$$

In order to obtain the potential (or force) at the points $\{y_j\}$ due to the charges at the points $\{x_i\}$ directly, we could compute

$$\sum_{i=1}^{m} \phi_{x_i}(y_j) \qquad \text{for all } j = 1, ..., n. \tag{2.15}$$

This clearly requires order nm work (evaluating m fields at n points). Now suppose that we first compute the coefficients of a p-term multipole expansion of the potential due to the charges $q_1, q_2, ..., q_m$ about x_0, using Theorem 2.1.1. This requires a number of operations proportional to mp. Evaluating the resulting multipole expansion at all points y_j requires order np work, and the total amount of computation is of the order $O(mp + np)$. Moreover, by (2.14),

$$\left| \sum_{i=1}^{m} \phi_{x_i}(y_j) - Q \log(y_j - x_0) - \sum_{k=1}^{p} \frac{a_k}{|y_j - x_0|^k} \right| \leq A \left(\frac{1}{2} \right)^p, \tag{2.16}$$

and in order to obtain a relative precision ϵ (with respect to the total charge), p must be of the order $-\log_2(\epsilon)$. Once the precision is specified, the amount of computation has been reduced to

$$O(m) + O(n), \tag{2.17}$$

which is significantly smaller than nm for large n and m.

2.2 Translation Operators and Error Bounds

The following three lemmas constitute the principal analytical tools of this chapter. Lemma 2.2.1 provides a formula for shifting the center of a multipole expansion, Lemma 2.2.2 describes how to convert such an expansion into a local (Taylor) expansion in a circular region of analyticity, and Lemma 2.2.3 furnishes a mechanism for shifting the center of a Taylor expansion within a region of analyticity. We also derive error bounds associated with these translation operators which allow us to carry out numerical computations to any specified accuracy.

LEMMA 2.2.1 (Translation of a Multipole Expansion) *Suppose that*

$$\phi(z) = a_0 \log(z - z_0) + \sum_{k=1}^{\infty} \frac{a_k}{(z - z_0)^k} \tag{2.18}$$

is a multipole expansion of the potential due to a set of m charges of strengths $q_1, q_2, \ldots, q_m$, all of which are located inside the circle D of radius R with center at z_0. Then for z outside the circle D_1 of radius $(R + |z_0|)$ and center at the origin,

$$\phi(z) = a_0 \log(z) + \sum_{l=1}^{\infty} \frac{b_l}{z^l}, \tag{2.19}$$

where

$$b_l = -\frac{a_0 z_0^l}{l} + \sum_{k=1}^{l} a_k z_0^{l-k} \binom{l-1}{k-1}, \tag{2.20}$$

with $\binom{l}{k}$ *the binomial coefficients. Furthermore, for any* $p \geq 1$,

$$\left| \phi(z) - a_0 \log(z) - \sum_{l=1}^{p} \frac{b_l}{z^l} \right| \leq \left(\frac{A}{1 - \left| \frac{|z_0| + R}{z} \right|} \right) \left| \frac{|z_0| + R}{z} \right|^{p+1} \tag{2.21}$$

with A defined in (2.11).

Proof: The coefficients of the shifted expansion (2.19) are obtained by expanding into a Taylor series the expression (2.18) with respect to z_0. For the error bound (2.21), observe that the terms $\{b_l\}$ are the coefficients of the (unique) multipole expansion about the origin of those charges contained in the circle D, and Theorem 2.1.1 applies immediately with r replaced by $|z_0| + R$.

Remark: Once the values $\{a_0, a_1, \ldots, a_p\}$ in the expansion (2.18) about z_0 are computed, we can obtain $\{b_1, \ldots, b_p\}$ exactly by (2.20). In other words, we may shift the center of a truncated multipole expansion without any loss of precision.

LEMMA 2.2.2(Conversion of a Multipole Expansion into a Local Expansion) *Suppose that m charges of strengths $q_1, q_2, \ldots, q_m$ are located inside the circle D_1 with radius R and center at z_0, and that $|z_0| > (c + 1)R$ with $c > 1$. (Figure 2.2.) Then the corresponding multipole expansion (2.18) converges inside the circle D_2 of radius R centered about the origin. Inside D_2, the potential due to the charges is described by a power series:*

$$\phi(z) = \sum_{l=0}^{\infty} b_l \cdot z^l, \tag{2.22}$$

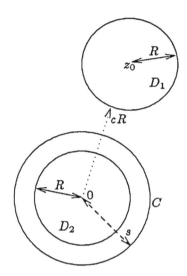

Figure 2.2
Source charges $q_1, q_2, ..., q_l$ are contained in the circle D_1. The corresponding multipole expansion about z_0 converges inside D_2. C is a circle of radius s, with $s > R$.

where

$$b_0 = a_0 \log(-z_0) + \sum_{k=1}^{\infty} \frac{a_k}{z_0^k}(-1)^k, \tag{2.23}$$

and

$$b_l = -\frac{a_0}{l \cdot z_0^l} + \frac{1}{z_0^l} \sum_{k=1}^{\infty} \frac{a_k}{z_0^k} \binom{l+k-1}{k-1}(-1)^k, \qquad \textit{for } l \geq 1. \tag{2.24}$$

Furthermore, for any $p \geq max\left(2, \frac{2c}{c-1}\right)$, an error bound for the truncated series is given by

$$\left| \phi(z) - \sum_{l=0}^{p} b_l \cdot z^l \right| < \frac{A(4e(p+c)(c+1) + c^2)}{c(c-1)} \left(\frac{1}{c}\right)^{p+1} \tag{2.25}$$

where A is defined in (2.11) and e is the base of natural logarithms.

Proof: We obtain the coefficients of the local expansion (2.22) from MacLaurin's theorem applied to the multipole expansion (2.18). To derive the error bound (2.25), we let $\gamma_0 = a_0 \, log(-z_0)$, $\gamma_l = -(\frac{a_0}{l \cdot z_0^l})$ for $l \geq 1$, and $\beta_l = b_l - \gamma_l$ for $l \geq 0$. Then

$$\left| \phi(z) - \sum_{l=0}^{p} b_l \cdot z^l \right| = \left| \sum_{l=p+1}^{\infty} b_l \cdot z^l \right| \leq S_1 + S_2 \qquad (2.26)$$

with

$$S_1 = \left| \sum_{l=p+1}^{\infty} \gamma_l \cdot z^l \right| \quad \text{and} \quad S_2 = \left| \sum_{l=p+1}^{\infty} \beta_l \cdot z^l \right|. \qquad (2.27)$$

A bound for S_1 is easily found by observing that

$$S_1 = \left| \sum_{l=p+1}^{\infty} \gamma_l \, z^l \right| \leq |a_0| \sum_{l=p+1}^{\infty} \frac{z^l}{l \cdot z_0^l} \leq A \sum_{l=p+1}^{\infty} \frac{z^l}{l \cdot z_0^l} \qquad (2.28)$$

$$\leq A \sum_{l=p+1}^{\infty} \left(\frac{1}{c+1} \right)^l < A \sum_{l=p+1}^{\infty} \left(\frac{1}{c} \right)^l = \left(\frac{A}{c-1} \right) \left(\frac{1}{c} \right)^p. \qquad (2.29)$$

To obtain a bound for S_2, let C be a circle of radius s where $s = cR \left(\frac{p-1}{p} \right)$ (Figure 2.2). Note first that for any $p \geq \frac{2c}{c-1}$,

$$R < \frac{cR + R}{2} < s < cR. \qquad (2.30)$$

Defining the function $\phi_1 : \mathbb{C} \setminus D_1 \to \mathbb{C}$ by the expression

$$\phi_1(z) = \phi(z) - a_0 \cdot log(z - z_0), \qquad (2.31)$$

and using Taylor's theorem for complex analytic functions (see [32], p. 190), we obtain

$$S_2 = \left| \phi_1(z) - \sum_{l=0}^{p} \beta_l \, z^l \right| = \left| \sum_{l=p+1}^{\infty} \beta_l \, z^l \right| \leq \frac{M}{1 - \frac{|z|}{s}} \left(\frac{|z|}{s} \right)^{p+1}, \qquad (2.32)$$

where

$$M = \max_{C} |\phi_1(t)|. \qquad (2.33)$$

Obviously, for any t lying on C,

$$|\phi_1(t)| \leq \sum_{k=1}^{\infty} \left| \frac{a_k}{(t-z_0)^k} \right|, \tag{2.34}$$

and it is easy to see that

$$|a_k| \leq AR^k \qquad \text{and} \qquad |t-z_0| \geq R + cR - s = R + \frac{cR}{p}. \tag{2.35}$$

After some algebraic manipulation, we have

$$M \leq A \left(\frac{pR+cR}{cR} \right), \qquad \text{and} \qquad 1 - \frac{|z|}{s} \geq \frac{cR-R}{cR+R}. \tag{2.36}$$

Observing that for any positive integer n and any integer $p \geq 2$,

$$\left(1 + \frac{1}{n}\right)^n \leq e \qquad \text{and} \qquad \left(1 + \frac{1}{p-1}\right)^2 \leq 4, \tag{2.37}$$

we obtain

$$S_2 \quad \leq \quad \frac{A(pR+cR)(cR+R)}{cR(cR-R)} \left(\frac{|z|}{cR} \right)^{p+1} \left(\frac{p}{p-1} \right)^{p+1} \tag{2.38}$$

$$\leq \quad \frac{A(p+c)(c+1)}{c(c-1)} \left(\frac{1}{c} \right)^{p+1} \left(1 + \frac{1}{p-1} \right)^{p-1} \left(1 + \frac{1}{p-1} \right)^2 \tag{2.39}$$

$$\leq \quad \frac{4Ae(p+c)(c+1)}{c(c-1)} \left(\frac{1}{c} \right)^{p+1} \tag{2.40}$$

Adding the last expression to the error bound for S_1 completes the proof.

The following lemma is an immediate consequence of MacLaurin's theorem. It describes an exact translation operation with a finite number of terms, and no error bound is needed.

LEMMA 2.2.3 (**Translation of a Local Expansion**) *For any complex z_0, z and $\{a_k\}$, $k = 0, 1, 2, \ldots, n$,*

$$\sum_{k=0}^{n} a_k (z-z_0)^k = \sum_{l=0}^{n} \left(\sum_{k=l}^{n} a_k \binom{k}{l} (-z_0)^{k-l} \right) z^l \tag{2.41}$$

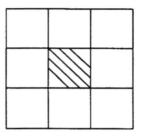

Figure 2.3
The computational box (shaded) and its nearest periodic images. The box is
centered at the origin and has area one.

2.3 The Fast Multipole Algorithm

In this section, we present an algorithm for the rapid evaluation of the potentials
and/or electrostatic fields due to distributions of charges. The central strategy used
is that of clustering particles at various spatial lengths and computing interactions
with other clusters which are sufficiently far away by means of multipole expansions.
Interactions with particles which are nearby are handled directly.

To be more specific, let us consider the geometry of the computational box,
depicted in Figure 2.3. It is a square with sides of length one, centered about the
origin of the coordinate system, and is assumed to contain all N particles of the
system under consideration. The eight nearest neighbor boxes are also shown, and
will be needed in the next section when considering various boundary conditions.
First, we will describe the method for free-space problems, where the boundary can
be ignored, and the only interactions to be accounted for involve particles within
the computational box itself.

We proceed by introducing a hierarchy of meshes which refine the computational
box into smaller and smaller regions (Figure 2.4). Mesh level 0 is equivalent to
the entire box, while mesh level $l + 1$ is obtained from level l by subdivision of
each region into four equal parts. The number of distinct boxes at mesh level l is
equal to 4^l. A tree structure is imposed on this mesh hierarchy, so that if *ibox* is

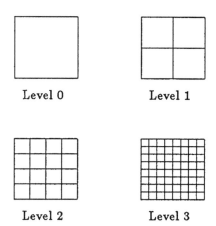

Figure 2.4
The computational box and three levels of refinement.

a fixed box at level l, the four boxes at level $l + 1$ obtained by subdivision of *ibox* are considered its children. Two boxes A and B, with sides of length $2s$, are said to be well-separated if they are separated by a distance $2s$. Let D_A and D_B be the smallest disks containing the boxes A and B, respectively. Then the disks have radii $\sqrt{2} \cdot s$, and the distance from the center of one disk to the closest point in the other disk is at least $(4 - \sqrt{2}) \cdot s$. Letting $c = (4 - \sqrt{2})/\sqrt{2} \approx 1.828$, the error bounds (2.10),(2.21) and (2.25) apply with a truncation error using p-term expansions of the order c^{-p}. Fixing a precision ϵ, we therefore choose $p = \lceil -\log_c(\epsilon) \rceil$ and specify that interactions only be computed by means of expansions for clusters of particles which are contained in well-separated boxes. This is precisely the condition needed for the desired precision to be achieved.

Other notation used in the description of the algorithm includes

$\Phi_{l,i}$ the p-term multipole expansion about the center of box i at level l , describing the far field potential due to the particles contained inside the box,

$\Psi_{l,i}$ the p-term local expansion about the center of box i at level l, describing the potential field due to all particles outside the box and its nearest neighbors,

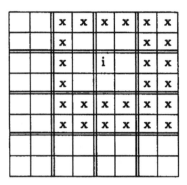

Figure 2.5
Interaction list for box i. Double lines correspond to mesh level 2 and thin lines to
level 3. Boxes marked with an "x" are well-separated from box i, and contained
within the nearest neighbors of box i's parent.

$\tilde{\Psi}_{l,i}$ the p-term local expansion about the center of box i at level l, describing the
potential field due to all particles outside i's *parent* box and the *parent* box's
nearest neighbors.

Interaction list: for box i at level l, it is the set of boxes which are children of
the nearest neighbors of i's parent and which are well-separated from box i
(Figure 2.5).

The fast multipole algorithm is a two-pass procedure. In the first (upward)
pass, we form the multipole expansions $\Phi_{l,i}$ for all boxes at all levels, beginning at
the finest level of refinement. In the second (downward) pass, we form the local
expansions $\Psi_{l,i}$ for all boxes at all levels, beginning at the coarsest level.

Suppose now that at level $l-1$, the local expansion $\Psi_{l-1,i}$ has been obtained for
all boxes. Then, by using lemma 2.2.3 to shift (for all i) the expansion $\Psi_{l-1,i}$ to
each of box i's children , we have, for each box j at level l, a local representation
of the potential due to all particles outside of j's parent's neighbors, namely $\tilde{\Psi}_{l,j}$.
The interaction list is, therefore, precisely that set of boxes whose contribution
to the potential must be added to $\tilde{\Psi}_{l,j}$ in order to create $\Psi_{l,j}$. This is done by
using lemma 2.2.2 to convert the multipole expansions of these interaction boxes to

local expansions about the current box center and adding them to the expansion obtained from the parent. Note also that with free-space boundary conditions, $\Psi_{0,i}$ and $\Psi_{1,i}$ are equal to zero since there are no well-separated boxes to consider, and we can begin forming local expansions at level 2.

Finally, for each box i at the finest level n, we evaluate the local expansion $\Psi_{n,i}$ for each particle contained in box i. It remains only to include the interactions of each particle in box i with those particles contained in box i and its nearest neighbors. These interactions are computed directly.

Following is a formal description of the algorithm.

Algorithm

Initialization

Choose a level of refinement $n \approx \lceil \log_4 N \rceil$, a precision ϵ, and set $p = \lceil -\log_c(\epsilon) \rceil$.

Upward Pass

Step 1

Comment [Form multipole expansions of potential field due to particles
in each box about the box center at the finest mesh level.]

do $ibox = 1, ..., 4^n$
    Form a p-term multipole expansion $\Phi_{n,ibox}$, by using Theorem 2.1.1.
enddo

Step 2

Comment [Form multipole expansions about the centers of all boxes
at all coarser mesh levels, each expansion representing the potential
field due to all particles contained in one box.]

do $l = n-1, ..., 0$
    do $ibox = 1, ..., 4^l$
        Form a p-term multipole expansion $\Phi_{l,ibox}$, by using
        lemma 2.2.1 to shift the center of each child box's expansion
        to the current box center and adding them together.
    enddo
enddo

Downward Pass

Comment [In the downward pass, interactions are consistently computed
        at the coarsest possible level. For a given box, this is accomplished
        by including interactions with those boxes which are well-separated
        and whose interactions have not been accounted for at the parent's
        level.]

Step 3

Comment [Form a local expansion about the center of each box at each mesh
        level $l \leq n - 1$. This local expansion describes the field due to all
        particles in the system that are not contained in the current box or
        its nearest neighbors. Once the local expansion is obtained for a given
        box, it is shifted, in the second inner loop to the centers of the box's
        children, forming the initial expansion for the boxes at the next
        level.]

Set $\tilde{\Psi}_{1,1} = \tilde{\Psi}_{1,2} = \tilde{\Psi}_{1,3} = \tilde{\Psi}_{1,4} = (0, 0, ..., 0)$
do $l = 1, ..., n - 1$
    do $ibox = 1, ..., 4^l$
        Form $\Psi_{l,ibox}$ by using lemma 2.2.2 to convert the multipole expansion
        $\Phi_{l,j}$ of each box j in *interaction list* of box *ibox*
        to a local expansion about the center of box *ibox*, adding these
        local expansions together, and adding the result to $\tilde{\Psi}_{l,ibox}$.
    enddo

    do $ibox = 1, ..., 4^l$
        Form the expansion $\tilde{\Psi}_{l+1,j}$ for *ibox*'s children by using
        Lemma 2.2.3 to expand $\Psi_{l,ibox}$ about the children's box centers.
    enddo
enddo

Step 4

Comment [Compute interactions at finest mesh level]

do $ibox = 1, ..., 4^n$
    Form $\Psi_{n,ibox}$ by using lemma 2.2.2 to convert the multipole
    expansion $\Phi_{n,j}$ of each box j in *interaction list* of box *ibox*
    to a local expansion about the center of box *ibox*, adding these
    local expansions together, and adding the result to $\tilde{\Psi}_{n,ibox}$.
enddo

Comment [Local expansions at finest mesh level are now available.
        They can be used to generate the potential or force due to all
        particles outside the nearest neighbor boxes at finest mesh level.]

Step 5

Comment [Evaluate local expansions at particle positions to obtain
the potential (or force) due to distant particles.]

do $ibox = 1, ..., 4^n$
    For every particle p_j located at the point z_j in box $ibox$,
    evaluate $\Psi_{n,ibox}(z_j)$.
enddo

Step 6

Comment [Compute potential (or force) due to nearest neighbors directly.]

do $ibox = 1, ..., 4^n$
    For every particle p_j in box $ibox$, directly compute interactions
    with all other particles within the box and its nearest neighbors.
enddo

Step 7

do $ibox = 1, ..., 4^n$
    For every particle in box $ibox$, add direct and far-field terms together.
enddo

Remark: Each local expansion is described by the coefficients of a p-term polynomial. Direct evaluation of this polynomial at a point yields the potential. But,
by lemma 2.1.1, the force is immediately obtained from the derivative which is
available analytically. There is no need for numerical differentiation. Furthermore,
due to the analyticity of Φ', there exist error bounds for the force of exactly the
same form as (2.10),(2.21) and (2.25).

A brief analysis of the algorithmic complexity is given below.

Step Number	Operation Count	Explanation
Step 1	order Np	each particle contributes to one expansion at the finest level.
Step 2	order Np^2	At the lth level, 4^l shifts involving order p^2 work per shift must be performed.
Step 3	order $\leq 28Np^2$	There are at most 27 entries in the interaction list for each box at each

		level. An extra order Np^2 work is required for the second loop.
Step 4	order $\leq 27Np^2$	Again, there are at most 27 entries in the interaction list for each box, and $\approx N$ boxes.
Step 5	order Np	One p-term expansion is evaluated for each particle.
Step 6	order $\frac{9}{2}Nk_n$	Let k_n be a bound on the number of particles per box at the finest mesh level. Interactions must be computed within the box and its eight nearest neighbors, but using Newton's third law, we need only compute half of the pairwise interactions.
Step 7	order N	Adding two terms for each particle.

The estimate for the running time is therefore

$$N \left(-2 \cdot a \cdot p + 56 \cdot b \cdot p^2 + 4.5 \cdot d \cdot k_n + e\right) , \tag{2.42}$$

with the constants $a, b, c, d,$ and e determined by the computer system, language, implementation, etc.

Remark: Note that implicit in the complexity estimate is a condition of homogeneity, namely that the number of particles per box at the finest mesh level is bounded. Non-homogeneous distributions are discussed in section 2.5.

In addition to the asymptotic time complexity, asymptotic storage requirements are an important characteristic of a numerical procedure. The algorithm requires that $\Phi_{l,j}$ and $\Psi_{l,j}$ be stored, as well as the locations of the particles, their charges, and the results of the calculations (the potentials and/or electric fields). Since every box at every level has a pair of p-term expansions, Φ and Ψ, associated with it, and the lengths of all other storage arrays are proportional to N, it is easy to see that the asymptotic storage requirements of the algorithm are of the form

$$(\alpha + \beta\, p) \cdot N , \tag{2.43}$$

with the coefficients α and β determined, as above, by the computer system, language, implementation, etc.

2.4 Boundary Conditions

A variety of boundary conditions are used in particle simulations, including periodic boundary conditions, homogeneous Dirichlet or Neumann conditions, and several types of mixed conditions. The periodic case will be treated first in some detail. We then turn to the imposition of Dirichlet conditions, and end with a brief discussion of the other cases.

2.4.1 Periodic Boundary Conditions

We assume that the periodic particle model has no net charge, and begin by reconsidering the computational domain depicted in Figure 2.3. At the end of the upward pass of the algorithm, we have a net multipole expansion

$$\Phi_{0,1}(z) = \sum_{k=1}^{p} \frac{a_k}{z^k} \tag{2.44}$$

for the entire computational box. This is then the expansion for each of the periodic images of the box with respect to its own center. All of these images except for the ones depicted in Figure 2.3 are well-separated from the computational box itself, and the fields they induce inside the computational domain are accurately representable by a p-term local expansion where, as before, $p = \lceil -\log_c(\epsilon) \rceil$ is the number of terms needed to achieve a relative precision ϵ. This local representation, given by Lemma 2.2.2, can be written as

$$\Psi_{0,1} = \sum_{m=1}^{p} b_m \cdot z^m \tag{2.45}$$

with

$$b_m = \frac{1}{z_0^m} \sum_{k=1}^{p} \frac{a_k}{z_0^k} \binom{m+k-1}{k-1} (-1)^k, \quad \text{for} \quad m = 0, 1, ..., p, \tag{2.46}$$

where z_0 the center of the image box under consideration.

Remark: In certain problems (e.g. cosmology), the computational box obviously cannot satisfy the condition of no net charge (mass). This condition is necessary for the potential to be well-defined, since the logarithmic term becomes unbounded as $z_0 \to \infty$. Force calculations, however, may still be carried out. Indeed, using the notation of the algorithm, $\Phi_{l,i}, \Psi_{l,i}, \tilde{\Psi}_{l,i}$ are expansions of analytic functions representing the potential, so that their derivatives are also analytic functions (with the same regions of analyticity). Moreover, it is clear from Theorem 2.1.1 that the derivatives $\Phi'_{l,i}$ are described by pure inverse power series. Therefore, the identical formal structure of the algorithm can, due to Lemma 2.1.1, be used to evaluate force fields everywhere, bypassing the difficulty introduced by the logarithmic term. The only change required is that the initial expansions computed be the derivatives of the multipole expansions and not the multipole expansions themselves.

Note now that well-separated images of the computational cell are boxes whose centers z_0 have integer real and imaginary parts, with $Re(z_0) \geq 2$ or $Im(z_0) \geq 2$. Let S be the set of such centers. To account for the field due to all well-separated images, we form the coefficients for the local representation by adding the local shifted expansions of the form (2.46) for all $z_0 \in S$ to obtain

$$b_m^{total} = \sum_{k=1}^{p} a_k \binom{m+k-1}{k-1} (-1)^k \left(\sum_S \frac{1}{z_0^{m+k}} \right). \tag{2.47}$$

Remark: Note that the total number of image boxes in S is infinite, and Lemma 2.2.2 is not directly applicable in this case. However, by combining (2.25) with the triangle inequality, it is easy to see that an estimate of the form (2.25) does indeed apply.

The summation over S for each inverse power of z_0 can be precomputed and stored. For $(m+k) > 2$, the series is absolutely convergent. However, for $(m+k) \leq 2$, the series is not absolutely convergent, and the computed value depends on the order of addition. Choosing a reasonable value for the sum of the series requires careful consideration of the physical model.

Suppose first that the only particle in the simulation is a charge of unit strength located at the origin. Then the periodic model corresponds to a uniform lattice of charges, and Newton's third law requires that the net force on each particle be zero. But the net force on the particle at the origin corresponds to the summation over S of $1/z_0$, so that we set

$$\sum_S \frac{1}{z_0} = 0. \tag{2.48}$$

To determine a value for the second term,

$$\sum_{S} \frac{1}{z_0^2} \,, \qquad (2.49)$$

suppose that the only particle in the simulation is a dipole of strength one, oriented along the x-axis and located at the origin. Then the periodic model is again a uniform lattice and the difference in potential between the equivalent sites $(-\frac{1}{2}, 0)$ and $(\frac{1}{2}, 0)$ must be zero; i.e.

$$\Phi_{(\frac{1}{2},0)} - \Phi_{(-\frac{1}{2},0)} \equiv \delta\Phi = 0. \qquad (2.50)$$

The contribution to the potential difference, $\delta\Phi$, of a single dipole located at z_0 is

$$\frac{1}{z_0 - \frac{1}{2}} - \frac{1}{z_0 + \frac{1}{2}} = \frac{1}{z_0^2 - \frac{1}{4}} \,. \qquad (2.51)$$

Thus, we find that the potential difference due to the original dipole located at the origin is -4. For an image dipole located at z_0, with $|z_0| \geq 1$, we can expand the contribution to $\delta\Phi$ as follows:

$$\frac{1}{z_0^2 - \frac{1}{4}} = \frac{1}{z_0^2} + \frac{1}{4z_0^4 - z_0^2}. \qquad (2.52)$$

Now let S' be the set of the centers of all image boxes. That is, S' is the set of all points z_0 with integer real and imaginary parts, excluding the origin. Then

$$\delta\Phi = -4 + \sum_{S'} \frac{1}{z_0^2} + \sum_{S'} \frac{1}{4z_0^4 - z_0^2} \,. \qquad (2.53)$$

A somewhat involved calculation shows that

$$\sum_{S'} \frac{1}{4z_0^4 - z_0^2} = 4 - \pi \,. \qquad (2.54)$$

Therefore, to satisfy (2.50), we set

$$\sum_{S'} \frac{1}{z_0^2} = \pi \,. \qquad (2.55)$$

Now

$$\sum_{S'} \frac{1}{z_0^2} = \sum_{S} \frac{1}{z_0^2} + \sum_{S'\backslash S} \frac{1}{z_0^2} \ , \tag{2.56}$$

and the sum $\sum_{S'\backslash S} \frac{1}{z_0^2}$ is easily evaluated and found to be equal to zero. Therefore, we have

$$\sum_{S} \frac{1}{z_0^2} = \pi \ , \tag{2.57}$$

and the summation over S for every inverse power of z_0 is defined.

The procedure of converting the multipole expansion of the whole computational cell $\Phi_{0,1}$ into a local expansion $\Psi_{0,1}$ which describes the potential field due to all well-separated images can be written, in the notation of the algorithm, as

$$\Psi_{0,1} = T \cdot \Phi_{0,1} \ , \tag{2.58}$$

where T is a constant p by p matrix whose entries are defined by the formula

$$T_{m,k} = \binom{m+k-1}{k-1}(-1)^k \left(\sum_{S} \frac{1}{z_0^{m+k}} \right) \ . \tag{2.59}$$

This can be viewed as the first step in the downward pass of the algorithm for periodic boundary conditions. At this point, we have accounted for all interactions excluding the ones within the immediate neighbors of the computational box as depicted in Figure 2.3. But the expansions $\Phi_{l,i}$ for boxes inside the computational cell are also the expansions of the corresponding boxes inside the nearest neighbor images of the computational cell. By adding to the interaction list the appropriate boxes, we maintain the formal structure of the algorithm and the associated computational complexity.

2.4.2   Dirichlet Boundary Conditions

We turn now to the imposition of homogeneous Dirichlet boundary conditions, namely

$$\Phi(x,y) = 0 \quad \text{for} \quad (x,y) \in \partial D \ , \tag{2.60}$$

where ∂D is the boundary of the computational domain. Analytically speaking, this can be accomplished by the method of images, described in detail below. In general terms, we consider the potential field to be composed of two parts; that is,

$$\Phi = \Phi_{sources} + \Phi_{images} , \qquad (2.61)$$

where $\Phi_{sources}$ is the field due to the particles inside the computational cell and Φ_{images} is the field due to selected image charges located outside the computational cell. The image charge positions and strengths are chosen so that

$$\Phi_{sources}(x, y) = -\Phi_{images}(x, y) \text{ for } (x, y) \in \partial D . \qquad (2.62)$$

For the computational domain we are considering, appropriate locations for the image charges can be determined by an iterative process, illustrated in Figure 2.6. We first reflect each particle p_i of charge strength σ_i in the computational cell across the top boundary line, and place an image charge of strength $-\sigma_i$ at that location, generating an image box which we denote $\bar{C}$ (Figure 2.6 (b)). The set of image charges is denoted by V_1, and the field they induce is called Φ_{V_1}. Adding Φ_{V_1} to $\Phi_{sources}$ clearly enforces the desired condition along the top boundary. To impose the boundary condition along the bottom of the computational cell, we must reflect all charges (source and image) currently in the model across the bottom boundary, generating two more image boxes (which are copies of C and $\bar{C}$). The set of all image charges after this second reflection step is denoted by V_2. Now, while $\Phi_{sources} + \Phi_{V_2}$ is equal to zero along the bottom boundary, the resulting field violates the top boundary condition. We therefore reflect again across the top boundary, creating two new image boxes and a new set of image charges V_3, such that $\Phi_{sources} + \Phi_{V_3}$ satisfies the top condition but violates the bottom one. By iterating in this manner, we generate a sequence of sets of image charges $\{V_i\}$ with

$$V_1 \subset V_2 \subset V_3 \subset \ldots \subset V \qquad (2.63)$$

where $V = \cup_{i=1}^{\infty} V_i$ is the set of charges contained in the infinite array of image boxes depicted in Figure 2.6 (c). It is easy to see that the nth reflection simply adds two image boxes (copies of C and $\bar{C}$) at a distance proportional to n, whose net contribution to the field inside the computational cell decays as $1/n$. The $(n + 1)^{st}$ reflection adds two such image boxes at a distance proportional to $n + 1$ in the opposite direction. Their net contribution decays as $1/n^2$. Therefore, the corresponding sequence of image fields $\{\Phi_{V_i}\}$ converges inside the computational

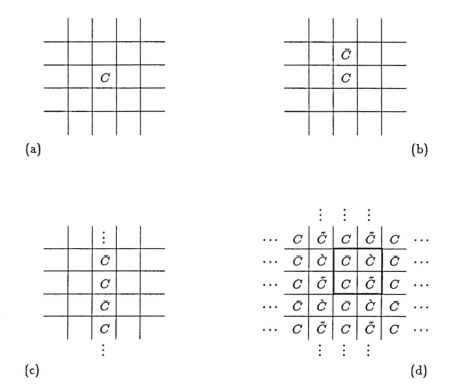

(a)   (b)

(c)   (d)

Figure 2.6
The computational cell centered at the origin is represented by C. $\bar{C}$, $\tilde{C}$, and $\grave{C}$ are images of C obtained by reflection across boundaries lines. See text for discussion.

cell, and the potential field $\Phi_{sources} + \Phi_V$ does satisfy both the top and bottom boundary conditions.

In order to enforce the Dirichlet condition on the remaining two sides, we proceed analogously. First, we reflect all the charges currently in the model (the original sources plus the images in V) across the left boundary. This obviously does not affect the top and bottom conditions, and enforces the homogeneous boundary condition along the left side of the computational cell. The current set of (all) image charges is now denoted H_1. Reflecting across the right boundary creates a new set H_2, with the field $\Phi_{sources} + \Phi_{H_2}$ satisfying the Dirichlet condition along the right (but not the left) boundary. Repeated reflection across the left and right boundaries of the computational cell yields a sequence $\{H_i\}$ of infinite sets of image charges,

$$H_1 \subset H_2 \subset H_3 \subset \ldots \subset H , \qquad (2.64)$$

where $H = \cup_{i=1}^{\infty} H_i$ is the set of charges contained in the two-dimensional family of image boxes depicted in Figure 2.6 (d). It is easy to see that the sequence $\{\Phi_{H_i}\}$ converges inside the computational cell, and we denote its limit by Φ_H. Finally, we observe that $\Phi_{sources} + \Phi_H = 0$ on the entire boundary ∂D.

From a computational point of view, the rate of convergence of the method of images is quite unsatisfactory. In conjunction with our algorithm, however, this method can be turned into an extremely efficient numerical tool. In the terminology previously introduced, all of the image boxes except the nearest neighbors of the computational cell are well-separated and their induced fields can be represented by a single local expansion, denoted $\Psi_{0,1}$. Once the coefficients of this local expansion have been computed, we need only account for interactions within the nearest neighbors of the computational cell itself. To do this, as in the periodic case, we simply add the appropriate image boxes to the interaction lists of the boxes inside the computational cell.

Thus, it remains only to calculate $\Psi_{0,1}$. We first observe that the plane of images has a periodic structure with unit "supercell" centered at $\left(\frac{1}{2}, \frac{1}{2}\right)$, indicated by thick lines in Figure 2.6 (d). But then, by the method developed above for periodic problems, we can obtain an expansion about the point $\left(\frac{1}{2}, \frac{1}{2}\right)$ which accounts for all interactions beyond the nearest neighbors of the supercell. This expansion can be converted, by using Lemma 2.2.3, into an expansion about the origin (the center of the computational cell), which we call $\tilde{\Psi}_{0,1}$. It remains to account for the well-separated boxes which are contained inside the supercell's nearest neighbors. There are exactly 27 of these boxes, and their multipole expansions can be shifted (by using Lemma 2.2.2) to local expansions about the origin which are then added to $\tilde{\Psi}_{0,1}$ to finally form $\Psi_{0,1}$.

2.4.3   Other Boundary Conditions

While in certain applications, periodic or Dirichlet boundary conditions are called for, in others, Neumann or mixed conditions have to be imposed on the boundary of the computational domain. A typical example of a problem with mixed conditions is the computational cell with Neumann conditions on two opposing sides and Dirichlet conditions on the two others. Other models require periodic boundary conditions on the left and right sides of the computational cell and Dirichlet or Neumann conditions on the top and bottom. The imposition of these conditions is achieved by a procedure essentially identical to the one described above. By reflection and/or periodic extension, one first generates an entire plane of images. The local expansion $\Psi_{0,1}$ is then computed by an appropriate summation over all well-separated image boxes, and the remaining image interactions are handled as above.

2.5   The adaptive algorithm

It is clear that for highly non-homogeneous distributions of particles, the algorithm of section 2.3 will perform poorly. We now describe an adaptive version of the fast multipcle algorithm whose running time is proportional to N, independent of the statistics of distribution. Since the extension of the method to problems with non-trivial boundary conditions is analogous to the process described in the previous section, we consider free-space problems only. The fundamental difference here is that we do not use the same number of levels for all parts of the computational box. Generally, this would result in a large number of empty boxes at finer levels of the procedure. Instead, some integer $s > 0$ is fixed, and at every level of refinement we subdivide only those boxes that contain more than s charges. At every level of refinement, a table of non-empty boxes is maintained, so that once an empty box is encountered, its existence is forgotten and it is completely ignored by the subsequent process.

Observation 2.5.1. It should be noted that for a fixed machine precision ϵ, only certain classes of particle distributions can be modeled, independent of the algorithm used. In particular, suppose that two charges c_1, c_2 in a distribution have positions x_1, x_2 and that $\|x_1 - x_2\| < \frac{\epsilon}{2} \cdot \|x_1 + x_2\|$. Obviously, under these conditions, the particles c_1 and c_2 can not be distinguished, and no meaningful

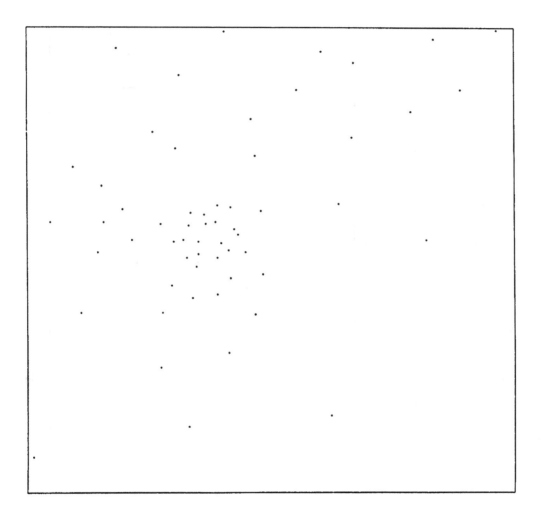

Figure 2.7
Non-uniform distribution of charges in the computational cell.

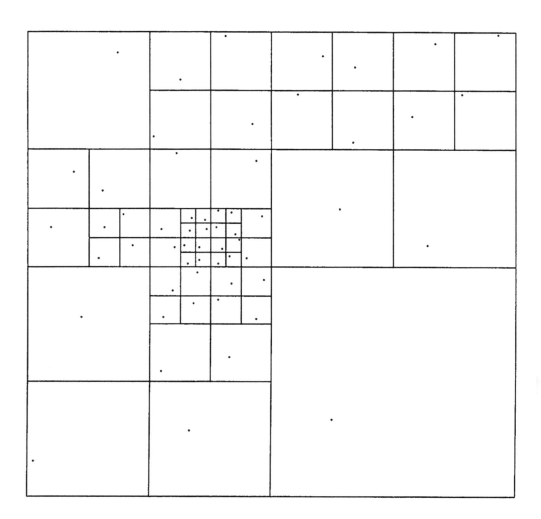

Figure 2.8
The hierarchy of meshes partitioning the computational cell.

simulation is possible. Since the smallest discernible distance between two particles depends on the actual positions of these particles in the computational cell, such a position-dependent condition can not be imposed a priori. In order to make the simulation possible, we will simply require that $r_{min} > \epsilon$, where r_{min} is the smallest distance between any two particles in the simulation, and ϵ is the machine precision. Therefore, the maximum number of ancestors for any box in the computational cell is $p = \lceil - \log_2(\epsilon) \rceil$.

Observation 2.5.2. If in Lemma 2.2.2, the field $\phi(z)$ is generated by a single charge located at z_0, then the only non-zero term in the expansion (2.18) is a_0, the charge strength.

2.5.1 Notation

In this subsection, we introduce several definitions to be used in the description of the algorithm below.

- For any subset A of the computational box, $T(A)$ will denote the set of particles that are contained in A.

- B_l is the set of non-empty boxes at level l. B_0 consists of only the computational box itself. We will denote by *nlev* the highest level of refinement at any point.

- The four boxes resulting from a box subdivision are referred to as *brothers*.

- If a box contains more than s particles, it is called a *parent box*. Otherwise, the box is said to be *childless*.

- A *child box* is a non-empty box resulting from the division into four of a parent box.

- *Colleagues* are adjacent boxes of the same size (at the same level). A given box has at most 8 colleagues (Figure 2.9).

With each box b at level l we will associate five lists of other boxes, determined by their positions with respect to b. Following are the definitions of these lists (Figure 2.10).

List 1 of a box b will be denoted by U_b; it is empty if b is a parent box. If b is childless, U_b consists of b and all childless boxes adjacent to b.

List 2 of a box b will be denoted V_b and is formed by all the children of the colleagues of b's parent that are well separated from b.

List 3 of a box b will be denoted by W_b. W_b is empty if b is a parent box, and consists of all descendants of b's colleagues whose parents are adjacent to b, but who are not adjacent to b themselves, if b is a childless box. Note that b is separated from each box w in W_b by a distance greater than or equal to the length of the side of w.

List 4 of a box b will be denoted by X_b and is formed by all boxes c such that $b \in W_c$. Note that all boxes in List 4 are childless and larger than b.

List 5 of a box b will be denoted by Y_b and consists of all boxes that are well separated from b's parent.

Finally,

Φ_b will denote the p-term multipole expansion about the center of b of the field created by all particles in $T(b)$.

Ψ_b will denote the p-term local expansion about the center of box b of the field created by all particles located outside $T(U_b) \cup T(W_b)$. $\Psi_b(r)$ is the result of the evaluation of the expansion Ψ_b at a particle r in $T(b)$.

Γ_b will denote the local expansion about the center of b of the field due to all particles in $T(V_b)$.

Δ_b will denote the local expansion about the center of b representing the field due to all charges located in $T(X_b)$.

$\alpha_b(r)$ will denote the the field at $r \in T(b)$ due to all particles in $T(U_b)$.

$\beta_b(r)$ will denote the field at $r \in T(b)$ due to all particles in $T(W_b)$.

2.5.2 Informal description of the algorithm

The algorithm can be viewed as a recursive process of subdividing the computational cell into increasingly finer meshes (see Figures 2.7 and 2.8). For a fixed box b at level l, the computational cell is partitioned into five subsets, U_b, V_b, W_b, X_b and Y_b, and the following procedure is applied to the sets of particles $T(U_b)$, $T(V_b)$, $T(W_b)$, $T(X_b)$ and $T(Y_b)$.

1. For each childless box b we combine the particles in $T(b)$ by means of Theorem 2.1.1 to form a multipole expansion Φ_b. For each parent box B we use Lemma 2.2.1 to merge the multipole expansions of its children into the expansion Φ_B.

2. The interactions between particles in $T(b)$ and $T(U_b)$ are computed directly. For each particle $r \in T(b)$, the result of these calculations is $\alpha_b(r)$.

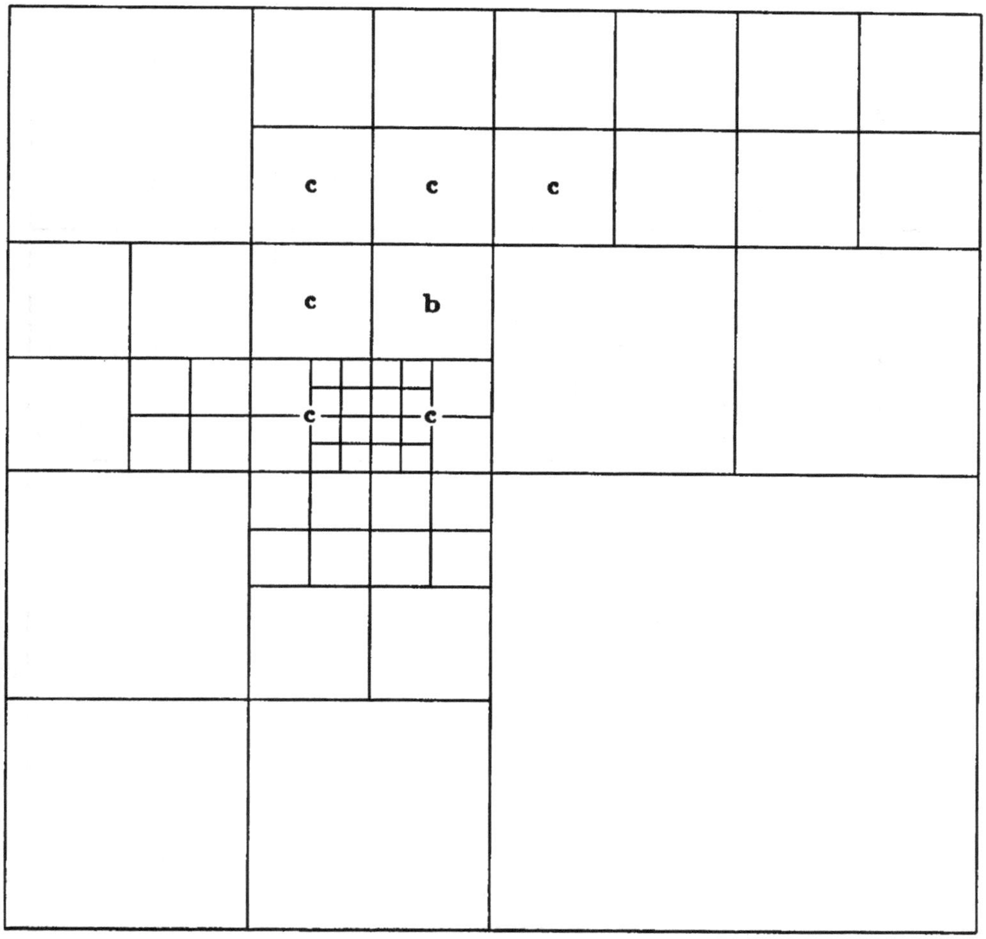

Figure 2.9
Box (b) and its colleagues (c).

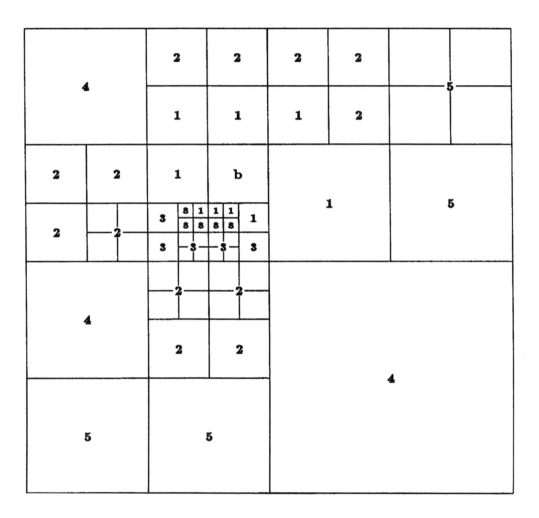

Figure 2.10
Box (b) and its associated lists 1 to 5 .

3. We use Lemma 2.2.2 to convert the multipole expansion of each box in V_b into a local expansion about the center of b, and add the resulting expansions to obtain Γ_b.

4. For every particle r in b, we compute the field $\beta_b(r)$ due to all particles in $T(W_b)$ by evaluating the p-term multipole expansions Φ_w of each box w in W_b at r, and adding them up.

5. We convert the field of each particle in $T(X_b)$ into a local expansion about the center of box b (see Observation 2.5.2), and add up the resulting expansions obtaining Δ_b.

6. We shift the center of the local expansion Γ_B of b's parent B to the centers of b and the other children of B by means of Lemma 2.2.3. We add the local expansion obtained to Γ_b.

7. For each box b, we evaluate the sum of the local expansions Γ_b and Δ_b at every particle r in b and add the result to $\alpha_b(r)$ and $\beta_b(r)$ obtaining the field at r.

Remark: Note that in the above procedure we never explicitly evaluate the interactions between particles in $T(b)$ and those in $T(Y_b)$. Indeed, since all boxes in Y_b are well separated from b's parent, the interaction between $T(Y_b)$ and $T(b)$ have been accounted for during steps 3 and 5 at a coarser level.

2.5.3   Formal description of the algorithm

Algorithm

Comment [Choose main parameters]

> Choose precision ϵ to be achieved. Set the number of terms
> in all expansions to $p = \lceil -\log_c(\epsilon) \rceil$. Choose the
> maximum number s of particles in a childless box.

Step 1.

Comment [Refine the computational cell into a hierarchy of meshes.]

```
do l = 1, 2, ···
    do b_i ∈ B_l
        if b_i contains more than s particles then
            subdivide b_i into four boxes, ignore the empty boxes
            formed, add the non-empty boxes formed to B_{l+1}.
        end if
    end do
end do
```

Comment [We denote by *nlev* the highest level of refinement, and by *nbox* the total number of boxes formed in Step 1.]

Step 2.

Comment [For every box b at every level l, form a multipole expansion representing the field outside b due to all the particles contained in b.]

Step 2.1

Comment [For each childless box b, use Theorem 2.1.1 to combine all charges inside b to obtain the multipole expansion about the center of b.]

```
do i=1,nbox
    if b_i is a childless box, use Theorem 2.1.1 to form a
    p-term expansion Φ_{b_i} representing the field
    outside b_i due to all charges located in b_i.
end do
```

Step 2.2

Comment [For each parent box b, use Lemma 2.2.1 to obtain the multipole expansion Φ_b by shifting the centers of the expansions of b's children to b's center, and adding the resulting expansions together.]

```
do l=nlev-1,1,-1
    do b_i ∈ B_l
        if b_i is a parent box then
            use Lemma 2.2.1 to shift the center of each of b_i's child
            box's expansion to b_i's center. Add the resulting expansions
            together to obtain the expansion Φ_{b_i}.
        end if
    end do
end do
```

Step 3.

Comment [For all particles in each childless box b, compute the interactions with all particles in $T(U_b)$ directly.]

> **do** i=1,nbox
>   **if** b_i is childless **then**
>   for each particle r in b_i, compute the sum $\alpha_b(r)$ of the interactions
>   between r and all particles in $T(U_{b_i})$.
>   **end if**
> **end do**

Step 4.

Comment [For each box b, use Lemma 2.2.2 to convert the multipole expansions of all boxes in V_b into local expansions about the center of box b.]

> **do** i=1,nbox
>   **do** $b_j \in V_{b_i}$
>   Convert multipole expansion Φ_{b_j} about b_j's center into
>   a local expansion about b_i's center using Lemma 2.2.2.
>   Add the resulting expansions to obtain Γ_{b_i}.
>   **end do**
> **end do**

Step 5.

Comment [For each childless box b, evaluate the multipole expansions of all boxes in W_b at every particle position in b.]

> **do** i=1,nbox
>   **if** b_i is childless **then**
>   Evaluate the multipole expansion Φ_{b_j} of each box $b_j \in W_{b_i}$
>   to obtain $\beta_{b_i}(r)$ for every particle r in box b_i.
>   **end if**
> **end do**

Step 6.

Comment [For each box b, use Lemma 2.2.2 and Observation 2.5.2 to form local expansions about the center b representing the field due to all particles in $T(X_b)$.]

```
do i=1,nbox
    Convert the field of every particle in T(X_{b_i}) into
    a local expansion about the center of b
end do
```

Step 7.

Comment [Use Lemma 2.2.3 to shift the centers of local expansions of parent boxes to the centers of their children.]

```
do l=1,nlev-1
    do b_i ∈ B_l
        if  b_i is a parent box then
            by using Lemma 2.2.3, shift the center of expansion Γ_{b_i}
            to the center of each of b_i's children b_j. Add
            the resulting expansion to Γ_{b_j}.
        end if
    end do
end do
```

Step 8.

Comment [For each childless box b, obtain Ψ_b as the sum of local expansions Γ_b and Δ_b. For each particle r in a childless box b, evaluate $\Psi_b(r)$ and obtain the field at r by adding $\Psi_b(r)$, $\alpha_b(r)$ and $\beta_b(r)$ together.]

```
do i=1,nbox
    if b_i is childless then
        Compute Ψ_{b_i} = Γ_{b_i} + Δ_{b_i}.
        For each particle r in b_i, evaluate Ψ_{b_i}(r).
        Add Ψ_{b_i}(r),α_{b_i}(r) and β_{b_i}(r) to obtain the field at r's position.
    end if
end do
```

An analysis of the complexity is given below.

Step number	Operation count	Explanation
Step 1	Np	Each particle is assigned to a box at every level. There are at most p levels of refinement.
Step 2.1	Np	Each particle contributes to the p-term expansion of one childless box.
Step 2.2	$\frac{5}{2}p^3N/s$	The center of the expansion of each box is shifted to the center of the parent box. The number of boxes is bounded by $5pN/s$ (see Lemma 2.6.5), and each shift requires $p^2/2$ work (see Lemma 2.2.1).
Step 3	$22pNs$	Each childless box b contains less than s particles and the work required to compute all interactions between particles in two boxes is $s^2/2$ when Newton's third law is used. The number of boxes in all List 1's is bounded by $44pN/s$ (see Lemmas 2.6.1 and 2.6.4).
Step 4	$80p^3N/s$	For each box, List 2 has no more than 32 entries (Lemma 2.6.2). There are at most $5pN/s$ boxes (Lemma 2.6.5) and each shift requires $p^2/2$ work (Lemma 2.2.2).
Step 5	$32p^2N$	Each childless box b contains less than s particles. The interactions of all particles in b and a box in W_b require ps work. The total number of boxes in List 3 is bounded by $32pN/s$ (Lemma 2.6.3 and 2.6.4).
Step 6	$32p^2N$	Each box in X_b contains fewer than s particles. The interactions between all particles in a box in X_b and box b require ps work. The total number of boxes in List 4 is bounded by $32pN/s$ (Lemma 2.6.3 and 2.6.4).
Step 7	$10p^3N/s$	Each box has at most four children. There are less than $5pN/s$ boxes (Lemma 2.6.5) and a shift requires $p^2/2$ work (Lemma 2.2.3).
Step 8	$Np + N$	A p-term expansion is evaluated at each particle position. The sums require an extra N work.

Summing up the contributions from all steps, we obtain the following CPU time estimate:

$$T = N \cdot (92.5ap^3/s + 64bp^2 + 22cps + 3dp + e), \qquad (2.65)$$

where the coefficients a, b, c, d, e depend on the computer system, language, implementation, etc. However, the parameter s (maximum permitted number of particles in a childless box) in (2.65) is not determined by the problem and can be choosen so as to minimize the resulting CPU time estimate. Differentiating (2.65) with respect to s, we obtain:

$$s_{min} = \sqrt{\frac{92.5a}{18c}} \cdot p \qquad (2.66)$$

and

$$T_{min} = N \cdot \left(\alpha p^2 + \beta p + \gamma \right) , \qquad (2.67)$$

with the constants α, β, γ determined by the computer system, language, implemantation, etc.

The storage requirements of the algorithm are determined by the number of non-empty boxes which is bounded by $5pN/s$. For each box we store the coefficients of a p-term multipole expansion and a p-term local expansion. The positions and charges of each particle also have to be stored. Therefore the storage requirements are of the form:

$$S = N \cdot \left(10fp/s + 3g \right) , \qquad (2.68)$$

where the coefficients f, g depend on the computer system, language, implementation, etc.

2.6   Algorithm Analysis

In this section, we prove several combinatorial lemmas that are used in Section 2.5 to estimate the complexity of the adaptive algorithm. We begin by introducing some additional notation.

Given a subdivision S of the computational cell and a childless box b in S, we will denote by S_b the subdivision obtained from S by subdividing b into 4 equal boxes, and refer to the process of obtaining S_b from S as an elementary refinement of S.

For a subdivision S of the computational cell, we will denote by B_S the set of all boxes in S.

C_S will denote the subset of B_S consisting of all childless boxes, i.e. boxes that are non-empty and not subdivided.

F_S will denote the subset of B_S consisting of all non-empty boxes.

D_S is the subset of B_S consisting of all empty boxes that have a childless brother.

$G_S = C_S \cup D_S$ is the subset of B_S consisting of all boxes b such that b is either childless or has a childless brother.

For any set of boxes A, $N(A)$ will denote the number of boxes in A.

LEMMA 2.6.1 *For any subdivision S of the computational cell*

$$\sum_{b \in C_S} N(U_b) \le 11 \cdot N(G_S). \tag{2.69}$$

Proof: We will prove the lemma by combining the following three assertions:

(a) Inequality (2.69) holds for the undivided computational cell.

(b) Any subdivision of the computational cell can be obtained by a sequence of elementary refinements of the computational cell.

(c) If an elementary refinement is applied to a subdivision for which (2.69) holds, it also holds in the refined subdivision.

The statements (a) and (b) above are obvious, and the following is a proof of (c).

Consider a subdivision S of the computational cell such that (2.69) is true for S, and a box b such that $b \in C_S$. Clearly

$$N(G_{S_b}) = N(G_S)) + 3, \tag{2.70}$$

and we will denote by U_b and U_b' the List 1's of b with respect to S and S_b, respectively. The following observations can be made about the List 1's of b and its children:

1. For any box $c \in C_S$, if $b \in U_c$ then $c \in U_b$.

2. Each child of b has itself and its three brothers in its List 1.

3. In the subdivision S_b, b is not childless and U'_b is empty.

4. Each box c in U_b is in the List 1 of at least one child of b.

5. The number of boxes of U_b that are in the List 1's of two children of b is bounded by 8.

It immediately follows from observations (1) - (5) above, that

$$\sum_{p \in C_{S_b}} N(U_p) - \sum_{q \in C_S} N(U_q) = 4 \cdot 4 + 2 \cdot [-N(U_b) + (N(U_b) + 8)] = 32, \qquad (2.71)$$

and combining (2.70) and (2.71) we obtain

$$\sum_{p \in C_{S_b}} N(U_p) \leq 11 \cdot N(G_{S_b}). \qquad (2.72)$$

LEMMA 2.6.2 *For any subdivision S of the computational cell*

$$\sum_{b \in F_S} N(V_b) \leq 32 \cdot N(F_S). \qquad (2.73)$$

Proof: Consider an arbitrary subdivision S of the computational cell, a box $c \in F_S$ and its parent box b. V_c is a subset of the children of b's colleagues, the maximum number of colleagues of c (or any other box) is eight, and each colleague can have four children. Therefore, the number of elements in V_c is bounded by 32.

LEMMA 2.6.3 *For any subdivision S of the computational cell*

$$\sum_{c \in C_S} N(W_c) = \sum_{b \in F_S} N(X_b) \leq 8 \cdot N(G_S). \qquad (2.74)$$

Proof: The first part of the Lemma is a direct consequence of the definition of List 4 (see Subsection 2.5.1): If a box b belongs W_c, then c belongs to X_b. Now, consider an arbitrary box $c \in F_S$, and its parent box b. The number of colleagues of b is certainly bounded by 8. We will denote by Z_b the set of all childless boxes which are adjacent to b and whose size is greater than or equal to that of b. The number of boxes in Z_b is bounded by 8, since each box in Z_b contains at least one of the eight colleagues of b, and no two such boxes can contain the same colleague. The second part of the lemma now follows from the observation that $W_c \subset Z_b$.

LEMMA 2.6.4 *For any subdivision S of the computational cell produced by the adaptive algorithm,*

$$N(C_S) \leq N(G_S) \leq 4 \cdot p \frac{N}{s}. \tag{2.75}$$

Proof: Each parent box b at level l contains more than s particles (otherwise, it would not have been subdivided any further). Therefore, the total number of parent boxes at level l is bounded by N/s. Each of these boxes can not have more than 4 children, and consequently the number of boxes in G_S at any level l is bounded by $4N/s$. Now, the conclusion of the lemma follows from Observation 2.5.1 and the obvious fact that $N(C_S) \leq N(G_S)$.

LEMMA 2.6.5 *For any subdivision S of the computational cell,*

$$N(F_S) \leq 5 \cdot p \cdot \frac{N}{s}. \tag{2.76}$$

Proof: The number of parent boxes at any level l is bounded by N/s, and each of them can not have more than 4 childless boxes at level $l+1$. Therefore, the sum of the numbers of non-empty boxes (all childless and parent boxes) at all levels is bounded by $p \cdot (N/s + 4N/s)$.

3 Potential Fields in Three Dimensions

Potential theory is frequently introduced in discussions of mathematical physics as the theory of Laplace's equation

$$\nabla^2 \Phi = \frac{\partial^2 \Phi}{\partial x^2} + \frac{\partial^2 \Phi}{\partial y^2} + \frac{\partial^2 \Phi}{\partial z^2} = 0 , \tag{3.1}$$

which arises in problems of gravitation, electrostatics, heat flow in homogeneous media, etc. In three dimensions, as in two dimensions, functions which satisfy the Laplace equation are referred to as harmonic functions.

Since we are primarily interested in the simulation of particle systems, we will continue to view potential theory as a means for studying the forces which are characterized by Newton's law of gravitation or Coulomb's law of electrostatic interaction. A thorough description of potential theory from this point of view is available in the classic text by Kellogg [25]. A shorter description is available in the text by Wallace [41], whose approach is followed somewhat more closely in this chapter.

For the sake of clarity, in the subsequent discussion we will always assume that we are faced with a problem in electrostatics. In other words, the physical system under consideration consists of a collection of charged particles with the potential energy and force obtained as the sum of pairwise interactions from Coulomb's law.

In Sections 3.1 to 3.3, we briefly develop the theory of spherical harmonics, while in Section 3.4, we describe the series expansion of the field due to an arbitrary distribution of charge. In Section 3.5, we present some new results concerning translation operators for the Laplace equation in three dimensions, and demonstrates how these operators can be used to manipulate both far field and local expansions in a manner which will be required by the fast multipole algorithm.

3.1   The field of a charge

Let a point charge of unit strength be located at the origin. Then, for any point $P = (x, y, z) \in \mathbb{R}^3$ with $\|P\| = r \neq 0$, the potential and electrostatic field due to this charge are described by the expressions

$$\Phi = \frac{1}{r} \tag{3.2}$$

and

$$\vec{E} = -\nabla \Phi = \left(\frac{x}{r^3}, \frac{y}{r^3}, \frac{z}{r^3} \right) , \tag{3.3}$$

respectively.

Suppose now that the unit charge is located at a point Q, not the origin. The potential at a point $P \neq Q$ is, of course, the inverse of the distance $PQ = r'$. We would like to derive a series expansion for the potential at P in terms of its distance from the origin r. To do this, we will use spherical coordinates with $P = (r, \theta, \phi)$ and $Q = (\rho, \alpha, \beta)$ (Fig. 3.1). Letting γ be the angle between the vectors P and Q, we have from the law of cosines

$$r'^2 = r^2 + \rho^2 - 2r\rho \cos \gamma \quad , \tag{3.4}$$

with

$$\cos \gamma = \cos \theta \cos \alpha + \sin \theta \sin \alpha \cos(\phi - \beta). \tag{3.5}$$

From this relation, we may write

$$\frac{1}{r'} = \frac{1}{r\sqrt{1 - 2\frac{\rho}{r}\cos \gamma + \frac{\rho^2}{r^2}}} = \frac{1}{r\sqrt{1 - 2u\mu + \mu^2}} \quad , \tag{3.6}$$

having set

$$\mu = \frac{\rho}{r} \quad \text{and} \quad u = \cos \gamma \ . \tag{3.7}$$

It can be shown that for $\mu < 1$, we may expand the inverse square root in powers of μ, resulting in a series of the form

$$\frac{1}{\sqrt{1 - 2u\mu + \mu^2}} = \sum_{n=0}^{\infty} P_n(u)\mu^n \tag{3.8}$$

where

$$P_0(u) = 1, \quad P_1(u) = u, \quad P_2(u) = \frac{3}{2}(u^2 - \frac{1}{3}), \cdots \tag{3.9}$$

and, in general, $P_n(u)$ is the Legendre polynomial of degree n. Our expression for the field now takes the form

$$\frac{1}{r'} = \sum_{n=0}^{\infty} \frac{\rho^n}{r^{n+1}} P_n(u). \tag{3.10}$$

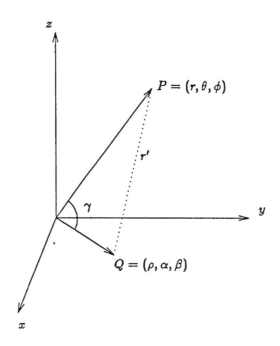

Figure 3.1
Points P and Q separated by a distance r', and subtending an angle γ between them.

Equation (3.10) is often referred to as a multipole expansion, and is said to describe the far field due to a charge at Q, since the condition for the validity of (3.8) is that $\mu < 1$ or $r > \rho$.

There is a duality inherent in the situation depicted in Figure 3.1, namely that if the locations of the charge (Q) and the evaluation point (P) were interchanged, then the field at P would still be described by $\frac{1}{r'}$. **In this case, so long as $r < \rho$, we may write**

$$\frac{1}{r'} = \frac{1}{\rho\sqrt{1 - 2\frac{r}{\rho}\cos\gamma + \frac{r^2}{\rho^2}}} = \sum_{n=0}^{\infty} \frac{r^n}{\rho^{n+1}} P_n(u). \tag{3.11}$$

Equation (3.11) is valid only in the open sphere centered at the origin with radius ρ, and we will refer to such a description of the potential field as a local expansion.

We turn now to an examination of the coefficients $P_n(u)$.

3.2   Legendre polynomials

The development of the field of a charge as a series is one of the many alternative ways of defining the Legendre polynomials, and is useful for studying some of their properties. For example, we have

LEMMA 3.2.1 $P_n(1) = 1$ *for $n = 0, 1, 2, \cdots$*

Proof :   Let $u = 1$. Then, using equation (3.8) above, we have

$$\sum_{n=0}^{\infty} \mu^n \cdot P_n(1) = \frac{1}{\sqrt{1 - 2\mu + \mu^2}} = \frac{1}{1 - \mu} = \sum_{n=0}^{\infty} \mu^n. \tag{3.12}$$

Equating the coefficients of successive powers of μ in the two series yields the desired result.

Two observations pertaining to Legendre polynomials will be needed below. Their proofs can be found in most standard textbooks (see, for example [25]).

LEMMA 3.2.2 *Let $u \in \mathbb{R}$, with $|u| \leq 1$. Then*

$$P_n(u) \leq 1. \tag{3.13}$$

LEMMA 3.2.3 *The Legendre polynomials may be written in the form*

$$P_n(u) = \sum_{k=0}^{\lfloor \frac{n}{2} \rfloor} \frac{1 \cdot 3 \cdots (2n - 2k - 1)}{2^k \cdot k!(n - 2k)!} \cdot (-1)^k u^{n-2k} \tag{3.14}$$

From Lemma 3.2.2, it is straightforward to obtain the following two error bounds.

LEMMA 3.2.4 *Suppose that a charge of strength q is located at the point $Q = (\rho, \alpha, \beta)$, and that $P = (r, \theta, \phi) \in R^3$, with $\|P - Q\| = r'$ and $r > \rho$. Letting γ be the angle between the two points, we have an error bound for the multipole expansion (3.10) of the form*

$$\left| \frac{q}{r'} - \sum_{n=0}^{p} \frac{q \cdot \rho^n}{r^{n+1}} P_n(\cos \gamma) \right| \leq \frac{q}{r - \rho} \left(\frac{\rho}{r} \right)^{p+1} . \tag{3.15}$$

Similarly, when $r < \rho$, we have an error bound for the local expansion (3.11) of the form

$$\left| \frac{q}{r'} - \sum_{n=0}^{p} \frac{q \cdot r^n}{\rho^{n+1}} P_n(\cos \gamma) \right| \leq \frac{q}{\rho - r} \left(\frac{r}{\rho} \right)^{p+1} . \tag{3.16}$$

Since the functions $P_n(u)$ arise in a series expansion of the field due to a charge, it is not surprising that they are related in some way to the partial derivatives of $\frac{1}{r}$. We will now make this relation more precise. Suppose, for simplicity, that the unit charge is located at the point $Q = (0, 0, \varsigma)$ on the z-axis, with $\varsigma > 0$. Then, the field at the point $P = (x, y, z)$ with spherical coordinates (r, θ, ϕ) is given by

$$\Phi = \frac{1}{PQ} = \frac{1}{\sqrt{x^2 + y^2 + (z - \varsigma)^2}} .$$

By expanding this expression for Φ into a Taylor series with respect to the variable ς, we obtain the formula

$$\Phi = \sum_{n=0}^{\infty} \frac{\varsigma^n}{n!} \left[\frac{\partial^n}{\partial \varsigma^n} \Phi \right]_{\varsigma=0}$$

Rewriting this expression in terms of derivatives with respect to z, we have

$$\Phi = \sum_{n=0}^{\infty} \frac{(-1)^n \varsigma^n}{n!} \frac{\partial^n}{\partial z^n} \left(\frac{1}{r} \right)$$

Comparing successive powers of ς in this expansion with the corresponding powers of ρ in equation (3.10) above, we find the following relation:

$$\frac{P_n(\cos\theta)}{r^{n+1}} = \frac{(-1)^n}{n!} \frac{\partial^n}{\partial z^n} \left(\frac{1}{r}\right) .$$

We note that since the partial derivatives of $\frac{1}{r}$ with respect to z must themselves be solutions of Laplace's equation, we have proved that the functions $P_n(\cos\theta)/r^{n+1}$ are harmonic.

Our analysis thus far allows us to develop as a series the far field potential due to point charges in two distinct settings. The first corresponds to the situation depicted in Figure 3.1 where the expansion (3.10) describes the field at a distance r from the origin due to a charge located at a distance ρ from the origin, with $\rho < r$. The resulting series, however, depends on the relative coordinates of the two particles. If another such series were developed for a second source at the point Q', they would have to be evaluated independently.

The second setting is one where a single series expansion can be obtained, describing the far field due to an entire collection of particles. For suppose that we are given m sources, all located on the z-axis at the points

$$\{(0,0,z_1),(0,0,z_2),\cdots,(0,0,z_m)\} , \tag{3.17}$$

with charge strengths $q_1, q_2, \cdots, q_m$. Then the field due to the ith charge at the point (r,θ,ϕ) is described by[1]

$$\Phi_i = \sum_{n=0}^{\infty} \frac{q_i z_i^n}{r^{n+1}} P_n(\cos\theta) \tag{3.18}$$

By the principle of superposition, we can add the coefficients of each of the charges' expansions together termwise, obtaining a power series describing the field due to all m sources, valid so long as $r > |z_i|$ for $i = 1, ..., m$. That is, the net potential is given by

$$\Phi = \sum_{n=0}^{\infty} \frac{\alpha_n}{r^{n+1}} P_n(\cos\theta) , \tag{3.19}$$

where

$$\alpha_n = \sum_{i=1}^{m} q_i z_i^n \tag{3.20}$$

---

[1] It is easy to verify that this result holds whether z_i is positive or negative by using lemma 3.2.3 to conclude that P_n is an even function when n is even and an odd function when n is odd.

In most problems of scientific interest, however, there is no restriction on the locations of the sources, and the preceding analysis is inapplicable. In the next section, we will investigate a more general approach to the solution of potential problems, which will allow us to compute asymptotic expansions of the field due to arbitrary distributions of charge.

3.3   Spherical Harmonics

The development of a general expansion describing potential fields in three dimensions is most clearly carried out by considering the Laplace equation itself, which characterizes the behavior of such fields in regions of free space. Using spherical coordinates, the Laplace equation (3.1) takes the form

$$\frac{1}{r^2}\frac{\partial}{\partial r}\left(r^2\frac{\partial\Phi}{\partial r}\right) + \frac{1}{r^2\sin\theta}\frac{\partial}{\partial\theta}\left(\sin\theta\frac{\partial\Phi}{\partial\theta}\right) + \frac{1}{r^2\sin^2\theta}\frac{\partial\,2\Phi}{\partial\phi^2} = 0. \qquad (3.21)$$

The standard solution of this equation by separation of variables results in an expression for the field as a series, the terms of which are known as spherical harmonics.

$$\Phi = \sum_{n=0}^{\infty}\sum_{m=-n}^{n}\left(L_n^m r^n + \frac{M_n^m}{r^{n+1}}\right)Y_n^m(\theta,\phi) \qquad (3.22)$$

In the above expansion, the terms $Y_n^m(\theta,\phi)r^n$ are usually referred to as spherical harmonics of degree n, the terms $Y_n^m(\theta,\phi)/r^{n+1}$ are called spherical harmonics of degree $-n-1$, and the coefficients L_n^m and M_n^m are known as the moments of the expansion.

Remark: It is obvious that in a far field (multipole) expansion, the coefficients L_n^m must be set to zero in order to satisfy the condition that the field decay at infinity. In a local expansion (which is to be analytic in a sphere centered at the origin), it is clearly the coefficients M_n^m which must be set to zero.

We noted previously that the functions $P_n(\cos\theta)/r^{n+1}$ are harmonic, having related them to the partial derivatives of $\frac{1}{r}$ with respect to z. But clearly the partial derivatives of $\frac{1}{r}$ with respect to x or y are also harmonic. The remainder of this section is devoted to describing the spherical harmonics of negative degree in terms of derivatives of $\frac{1}{r}$, and then to expressing the terms $Y_n^m(\theta, \phi)$ in a more computationally useful form.

Lemmas 3.3.1 - 3.3.3 below are well-known. Their proofs can be found, for example, in [23] or [41].

LEMMA 3.3.1

$$\frac{Y_n^0(\theta, \phi)}{r^{n+1}} = A_n^0 \cdot \frac{\partial^n}{\partial z^n} \left(\frac{1}{r} \right) . \tag{3.23}$$

For $m > 0$, we have

$$\frac{Y_n^m(\theta, \phi)}{r^{n+1}} = A_n^m \cdot (\frac{\partial}{\partial x} + i \frac{\partial}{\partial y})^m \left(\frac{\partial}{\partial z} \right)^{n-m} \left(\frac{1}{r} \right) , \tag{3.24}$$

and

$$\frac{Y_n^{-m}(\theta, \phi)}{r^{n+1}} = A_n^m \cdot (\frac{\partial}{\partial x} - i \frac{\partial}{\partial y})^m \left(\frac{\partial}{\partial z} \right)^{n-m} \left(\frac{1}{r} \right) , \tag{3.25}$$

where

$$A_n^m = \frac{(-1)^n}{\sqrt{(n-m)! \cdot (n+m)!}} . \tag{3.26}$$

Remark: The standard definition of the functions $Y_n^m(\theta, \phi)$ includes a normalization factor of

$$\sqrt{(2n+1)/4\pi}. \tag{3.27}$$

We will consistently use the definition given above. That is, the coefficient (3.27) will always be omitted.

Since certain differential operators arise frequently in discussions of spherical harmonics, we introduce the following notation.

Definition 3.3.1 *The operators ∂_+, ∂_-, and ∂_z are defined by the expressions*

$$\partial_\pm = \frac{\partial}{\partial x} \pm i \cdot \frac{\partial}{\partial y} \quad and \tag{3.28}$$

$$\partial_z = \frac{\partial}{\partial z} \tag{3.29}$$

LEMMA 3.3.2 *If ϕ is a harmonic function, then*

$$\partial_+ \partial_-(\phi) = -\partial_z^2(\phi) \tag{3.30}$$

LEMMA 3.3.3 *For any $n \geq m \geq 0$,*

$$\partial_{\pm}^m \partial_z^{n-m} \left(\frac{1}{r} \right) = (-1)^n (n-m)! \frac{1}{r^{n+1}} \cdot P_n^m(\cos\theta) \cdot e^{\pm im\phi} , \tag{3.31}$$

where the term $P_n^m(\cos\theta)$ denotes the associated Legendre function of degree n and order m.

Combining Lemmas 3.3.3 and 3.3.1, we have an expression for the spherical harmonics in terms of associated Legendre functions:

$$Y_n^m(\theta, \phi) = \sqrt{\frac{(n-|m|)!}{(n+|m|)!}} \cdot P_n^{|m|}(\cos\theta) e^{im\phi}. \tag{3.32}$$

We may then compute the function values $Y_n^m(\theta, \phi)$ by using the the following recursion relations [19].

$$(2n+1)x P_n^m(x) = (n-m+1)P_{n+1}^m(x) + (n+m)P_{n-1}^m , \tag{3.33}$$

and

$$P_n^{m+2}(x) + 2(m+1)\frac{x}{\sqrt{x^2-1}} P_n^{m+1}(x) = (n-m)(n+m+1)P_n^m(x) . \tag{3.34}$$

3.4 The Field Due to Arbitrary Distributions of Charge

We will need a well-known result from the theory of spherical harmonics, which is usually referred to as the Addition Theorem [41].

THEOREM 3.4.1 (Addition Theorem for Legendre Polynomials) *Let P and Q be points with spherical coordinates (r, θ, ϕ) and (ρ, α, β), respectively, and let γ be the angle subtended between them (Figure 3.1). Then*

$$P_n(\cos\gamma) = \sum_{m=-n}^{n} Y_n^{-m}(\alpha, \beta) \cdot Y_n^m(\theta, \phi) . \tag{3.35}$$

It is a straightforward matter now to form a multipole expansion describing the far field due to a collection of particles.

THEOREM 3.4.2(Multipole Expansion).   *Suppose that k charges of strengths $\{q_i, \ i = 1,...,k\}$ are located at the points $\{Q_i = (\rho_i, \alpha_i, \beta_i), \ i = 1,...,k\}$, with $|\rho_i| < a$. Then for any $P = (r, \theta, \phi) \in \mathbf{R}^3$ with $r > a$, the potential $\phi(P)$ is given by*

$$\phi(P) = \sum_{n=0}^{\infty} \sum_{m=-n}^{n} \frac{M_n^m}{r^{n+1}} \cdot Y_n^m(\theta, \phi) \ , \qquad (3.36)$$

where

$$M_n^m = \sum_{i=1}^{k} q_i \cdot \rho_i^n \cdot Y_n^{-m}(\alpha_i, \beta_i). \qquad (3.37)$$

Furthermore, for any $p \geq 1$,

$$\left| \phi(P) - \sum_{n=0}^{p} \sum_{m=-n}^{n} \frac{M_n^m}{r^{n+1}} \cdot Y_n^m(\theta, \phi) \right| \leq \frac{A}{r-a} \left(\frac{a}{r} \right)^{p+1} \ , \qquad (3.38)$$

where

$$A = \sum_{i=1}^{k} |q_i| \ . \qquad (3.39)$$

Proof :   Let us first consider the contribution from a single charge q_i located at $(\rho_i, \alpha_i, \beta_i)$. From formula (3.10) and the Addition Theorem for Legendre Polynomials, we have

$$\begin{aligned} \phi_i &= \sum_{n=0}^{\infty} \frac{q_i \cdot \rho_i^n}{r^{n+1}} \cdot P_n(\cos \gamma) \\ &= \sum_{n=0}^{\infty} \sum_{m=-n}^{n} \frac{[q_i \cdot \rho_i^n \cdot Y_n^{-m}(\alpha_i, \beta_i)]}{r^{n+1}} \cdot Y_n^m(\theta, \phi) \ . \end{aligned}$$

The coefficients M_n^m in equation (3.37) are then obtained by superposition. The error bound is an immediate consequence of (3.15), the triangle inequality, and the fact that the ratios ρ_i/r are bounded from above by a/r.

Before proceeding with the further development of the theory of spherical harmonics, we will demonstrate with a simple example how multipole expansions can

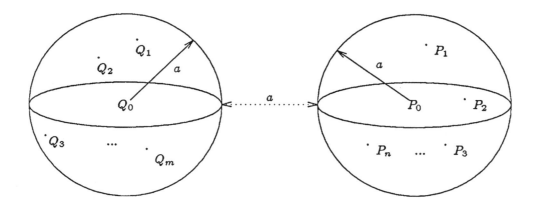

Figure 3.2
Well-separated sets in $\mathbb{R}^3$.

be used to reduce the computational complexity of the evaluation of potential fields. This is just the three-dimensional analog of the example given in Section 2.1. Suppose that a collection of k point charges of strengths $\{q_i, \; i = 1, ..., k\}$ are located at the points $\{Q_i = (\rho_i, \alpha_i, \beta_i), \; i = 1, \cdots, k\}$, and that $\{P_j = (r_j, \theta_j, \phi_j), \; j = 1, \cdots, n\}$ is another set of points in $\mathbb{R}^3$ (Figure 3.2). We say that the sets $\{Q_i\}$ and $\{P_j\}$ are *well-separated* if there exist points $P_0, Q_0 \in \mathbb{R}^3$ and a real number $a > 0$ such that

$$\|Q_i - Q_0\| \; < \; a \qquad \text{for } i = 1, ..., k \; ,$$
$$\|P_j - P_0\| \; < \; a \qquad \text{for } j = 1, ..., n \; , \quad \text{and}$$
$$\|Q_0 - P_0\| \; > \; 3a.$$

In order to obtain the potential at each of the points P_j due to the charges at the points Q_i directly, we could compute

$$\sum_{i=1}^{k} \phi_{Q_i}(P_j) \qquad \text{for } j = 1, ..., n. \tag{3.40}$$

This requires order $n \cdot k$ work (evaluating k fields at n points). Suppose, on the other hand, that we first compute the coefficients of a pth-degree multipole expansion of the potential due to the charges $q_1, q_2, ..., q_k$ about Q_0, using Theorem 3.4.2. This requires a number of operations proportional to $k \cdot p^2$. Evaluating the resulting multipole expansion at all points P_j requires order $n \cdot p^2$ work, and the total amount of computation is of the order $O(k \cdot p^2 + n \cdot p^2)$. Moreover, by (3.38),

$$\left| \sum_{i=1}^{k} \phi_{Q_i}(P_j) - \sum_{n=0}^{p} \sum_{m=-n}^{n} \frac{M_n^m}{\|P_j - Q_0\|^{n+1}} \cdot Y_n^m(\theta_j, \phi_j) \right| \leq \left(\frac{\sum_{i=1}^{k} |q_i|}{a} \right) \left(\frac{1}{2} \right)^{p+1} ,$$

and in order to obtain a relative precision ϵ (with respect to the total charge), p must be of the order $\lceil -\log_2(\epsilon) \rceil$. Once the precision is specified, the amount of computation has been reduced to

$$O(k) + O(n) , \tag{3.41}$$

which is a significant reduction in complexity when compared with the direct method.

3.5   Translation Operators and Error Bounds

As in the two-dimensional case, the principal analytical tools required by the fast algorithm are certain translation operators, acting on both multipole and local expansions. In order to develop the necessary formulae for these procedures, we will need the following three theorems, which can be viewed as generalizations of the classical addition theorem for Legendre polynomials. While a somewhat different form of Theorem 3.5.3 below can be found in the literature [10,14,38], Theorems 3.5.1 and 3.5.2 appear to be new. The following theorem describes a formula for the expansion of a spherical harmonic of negative degree about a shifted origin.

THEOREM 3.5.1(First Addition Theorem) *Let $Q = (\rho, \alpha, \beta)$ be the center of expansion of an arbitrary spherical harmonic of negative degree. Let the point $P =$*

(r, θ, ϕ), *with* $r > \rho$ *and* $P - Q = (r', \theta', \phi')$. *Then*

$$\frac{Y_{n'}^{m'}(\theta', \phi')}{r'^{\,n'+1}} = \sum_{n=0}^{\infty} \sum_{m=-n}^{n} \frac{J_m^{m'} \cdot A_n^m \cdot A_{n'}^{m'} \cdot \rho^n \cdot Y_n^{-m}(\alpha, \beta)}{A_{n+n'}^{m+m'}} \cdot \frac{Y_{n+n'}^{m+m'}(\theta, \phi)}{r^{n+n'+1}} \qquad (3.42)$$

where

$$J_m^{m'} = \begin{cases} (-1)^{min(|m'|, |m|)}, & \textit{if } m \cdot m' < 0; \\ 1, & \textit{otherwise.} \end{cases} \qquad (3.43)$$

Proof :   Making use of equation (3.10), the Addition Theorem for Legendre polynomials, and Lemma 3.3.1, we observe that

$$
\begin{aligned}
\frac{1}{\|P - Q\|} \;\; &= \;\; \frac{1}{r'} = \sum_{n=0}^{\infty} \frac{\rho^n}{r^{n+1}} \cdot P_n(\cos \gamma) \\
&= \;\; \sum_{n=0}^{\infty} \sum_{m=-n}^{n} [\rho^n \cdot Y_n^{-m}(\alpha, \beta)] \cdot \frac{Y_n^m(\theta, \phi)}{r^{n+1}} \\
&= \;\; \sum_{n=0}^{\infty} \Big(\sum_{m=-n}^{0} \rho^n \cdot Y_n^{-m}(\alpha, \beta) \cdot A_n^m \cdot \partial_-^{|m|} \partial_z^{n-|m|} \left(\frac{1}{r} \right) + \\
&\qquad\qquad \sum_{m=1}^{n} \rho^n \cdot Y_n^{-m}(\alpha, \beta) \cdot A_n^m \cdot \partial_+^m \partial_z^{n-m} \left(\frac{1}{r} \right) \Big) . \qquad (3.44)
\end{aligned}
$$

We now consider three separate cases.

Case I : $m' = 0$. From Lemma 3.3.1,

$$\frac{Y_{n'}^0(\theta', \phi')}{r'^{\,n'+1}} = A_{n'}^0 \cdot \partial_z^{n'} \left(\frac{1}{r'} \right) . \qquad (3.45)$$

Combining (3.44) and (3.45), we obtain

$$
\begin{aligned}
\frac{Y_{n'}^0(\theta', \phi')}{r'^{\,n'+1}} \;\; &= \;\; \sum_{n=0}^{\infty} \Big(\sum_{m=-n}^{0} \rho^n \cdot Y_n^{-m}(\alpha, \beta) \cdot A_{n'}^0 \cdot A_n^m \cdot \partial_-^{|m|} \partial_z^{n+n'-|m|} \left(\frac{1}{r} \right) + \\
&\qquad\qquad \sum_{m=1}^{n} \rho^n \cdot Y_n^{-m}(\alpha, \beta) \cdot A_{n'}^0 \cdot A_n^m \cdot \partial_+^m \partial_z^{n+n'-m} \left(\frac{1}{r} \right) \Big) \\
&= \;\; \sum_{n=0}^{\infty} \sum_{m=-n}^{n} \left(\frac{\rho^n \cdot Y_n^{-m}(\alpha, \beta) \cdot A_{n'}^0 \cdot A_n^m}{A_{n+n'}^m} \right) \cdot \frac{Y_{n+n'}^m(\theta, \phi)}{r^{n+n'+1}} ,
\end{aligned}
$$

where the last equality is obtained by another application of Lemma 3.3.1.

Case II : $m' < 0$. Using Lemma 3.3.1 again,

$$
\frac{Y_{n'}^{m'}(\theta', \phi')}{r'^{n'+1}}
$$

$$
= A_{n'}^{m'} \cdot \partial_-^{|m'|} \partial_z^{n'-|m'|} \left(\frac{1}{r'} \right)
$$

$$
= \sum_{n=0}^{\infty} \Big(\sum_{m=-n}^{0} \rho^n \cdot Y_n^{-m}(\alpha, \beta) \cdot A_{n'}^{m'} \cdot A_n^m \cdot \partial_-^{|m'|+|m|} \partial_z^{n+n'-|m|-|m'|} \left(\frac{1}{r} \right) +
$$

$$
\sum_{m=1}^{n} \rho^n \cdot Y_n^{-m}(\alpha, \beta) \cdot A_{n'}^{m'} \cdot A_n^m \cdot \partial_-^{|m'|} \partial_+^m \partial_z^{n+n'-m-|m'|} \left(\frac{1}{r} \right) \Big)
$$

$$
= \sum_{n=0}^{\infty} \sum_{m=-n}^{n} \left(\frac{J_m^{m'} \cdot A_{n'}^{m'} \cdot A_n^m \cdot \rho^n \cdot Y_n^{-m}(\alpha, \beta)}{A_{n+n'}^{m+m'}} \right) \cdot \frac{Y_{n+n'}^{m+m'}(\theta, \phi)}{r^{n+n'+1}} ,
$$

where

$$
J_m^{m'} = \begin{cases} 1, & \text{if } m \le 0; \\ (-1)^{min(|m'|,m)}, & \text{if } m > 0. \end{cases} \tag{3.46}
$$

To obtain the last equality, for the terms with $m > 0$, we have used Lemma 3.3.2 to annihilate whichever of the operators ∂_- and ∂_+ occurs less frequently.

Case III : $m' > 0$. From Lemma 3.3.1,

$$
\frac{Y_{n'}^{m'}(\theta', \phi')}{r'^{n'+1}}
$$

$$
= A_{n'}^{m'} \cdot \partial_+^{m'} \partial_z^{n'-m'} \left(\frac{1}{r'} \right)
$$

$$
= \sum_{n=0}^{\infty} \Big(\sum_{m=-n}^{0} \rho^n \cdot Y_n^{-m}(\alpha, \beta) \cdot A_{n'}^{m'} \cdot A_n^m \cdot \partial_+^{m'} \partial_-^{|m|} \partial_z^{n+n'-|m|-m'} \left(\frac{1}{r} \right) +
$$

$$
\sum_{m=1}^{n} \rho^n \cdot Y_n^{-m}(\alpha, \beta) \cdot A_{n'}^{m'} \cdot A_n^m \cdot \partial_+^{m+m'} \partial_z^{n+n'-m-m'} \left(\frac{1}{r} \right) \Big)
$$

$$
= \sum_{n=0}^{\infty} \sum_{m=-n}^{n} \left(\frac{J_m^{m'} \cdot A_{n'}^{m'} \cdot A_n^m \cdot \rho^n \cdot Y_n^{-m}(\alpha, \beta)}{A_{n+n'}^{m+m'}} \right) \cdot \frac{Y_{n+n'}^{m+m'}(\theta, \phi)}{r^{n+n'+1}} ,
$$

where

$$J_m^{m'} = \begin{cases} 1, & \text{if } m \geq 0; \\ (-1)^{min(m',|m|)}, & \text{if } m < 0. \end{cases} \qquad (3.47)$$

As before, for the terms with $m < 0$, we have used Lemma 3.3.2 to annihilate whichever of the operators ∂_- and ∂_+ occurs less frequently.

The second addition theorem yields a formula for converting a spherical harmonic of negative degree (a multipole term) with respect to one origin into a local expansion about a shifted origin.

THEOREM 3.5.2 (Second Addition Theorem) *Let* $Q = (\rho, \alpha, \beta)$ *be the center of expansion of an arbitrary spherical harmonic of negative degree. Let the point* $P = (r, \theta, \phi)$, *with* $r < \rho$ *and* $P - Q = (r', \theta', \phi')$. *Then*

$$\frac{Y_{n'}^{m'}(\theta', \phi')}{r'^{n'+1}} = \sum_{n=0}^{\infty} \sum_{m=-n}^{n} \frac{J_m^{m'} \cdot A_n^m \cdot A_{n'}^{m'} \cdot Y_{n+n'}^{m'-m}(\alpha, \beta)}{\rho^{n+n'+1} \cdot A_{n+n'}^{m-m'}} \cdot Y_n^m(\theta, \phi) r^n , \qquad (3.48)$$

where

$$J_m^{m'} = \begin{cases} (-1)^{n'}(-1)^{min(|m'|,|m|)}, & \text{if } m \cdot m' > 0; \\ (-1)^{n'}, & \text{otherwise.} \end{cases} \qquad (3.49)$$

Proof: We first let (x_P, y_P, z_P) and (x_Q, y_Q, z_Q) denote the Cartesian coordinates of the points P and Q, respectively. Then

$$\begin{aligned} \frac{\partial}{\partial x_P}\left(\frac{1}{r'}\right) &= -\frac{\partial}{\partial x_Q}\left(\frac{1}{r'}\right) \\ \frac{\partial}{\partial y_P}\left(\frac{1}{r'}\right) &= -\frac{\partial}{\partial y_Q}\left(\frac{1}{r'}\right) \\ \frac{\partial}{\partial z_P}\left(\frac{1}{r'}\right) &= -\frac{\partial}{\partial z_Q}\left(\frac{1}{r'}\right) \end{aligned} \qquad (3.50)$$

We will denote by $\partial_{+_P}, \partial_{-_P}, \partial_{z_P}, \partial_{+_Q}, \partial_{-_Q}, \partial_{z_Q}$, the differential operators given by Definition 3.3.1, with respect to the indicated variable point.

Combining equation (3.11), the Addition Theorem for Legendre Polynomials, and Lemma 3.3.1, we now obtain

$$
\begin{aligned}
\frac{1}{\|P - Q\|} = \frac{1}{r'} \;\; &= \;\; \sum_{n=0}^{\infty} \frac{r^n}{\rho^{n+1}} \cdot P_n(\cos\gamma) \\
&= \;\; \sum_{n=0}^{\infty} \sum_{m=-n}^{n} \frac{Y_n^{-m}(\alpha,\beta)}{\rho^{n+1}} \cdot Y_n^m(\theta,\phi) \cdot r^n \\
&= \;\; \sum_{n=0}^{\infty} \Big(\sum_{m=-n}^{0} A_n^m \cdot \partial_{+Q}^{|m|} \partial_{z_Q}^{n-|m|} \left(\frac{1}{\rho}\right) \cdot Y_n^m(\theta,\phi) \cdot r^n + \\
&\qquad \sum_{m=1}^{n} A_n^m \cdot \partial_{-Q}^{m} \partial_{z_Q}^{n-m} \left(\frac{1}{\rho}\right) \cdot Y_n^m(\theta,\phi) \cdot r^n \Big).
\end{aligned}
$$

Case I : $m' = 0$. Due to Lemma 3.3.1 and (3.50), we have

$$
\begin{aligned}
\frac{Y_{n'}^0(\theta',\phi')}{r'^{\,n'+1}} \\
= \;\; & A_{n'}^0 \cdot \partial_{z_P}^{n'} \left(\frac{1}{r'}\right) \\
= \;\; & \sum_{n=0}^{\infty} \Big(\sum_{m=-n}^{0} (-1)^{n'} A_{n'}^0 \cdot A_n^m \cdot \partial_{+Q}^{|m|} \partial_{z_Q}^{n+n'-|m|} \left(\frac{1}{\rho}\right) \cdot Y_n^m(\theta,\phi) \cdot r^n + \\
& \qquad \sum_{m=1}^{n} (-1)^{n'} A_{n'}^0 \cdot A_n^m \cdot \partial_{-Q}^{m} \partial_{z_Q}^{n+n'-m} \left(\frac{1}{\rho}\right) \cdot Y_n^m(\theta,\phi) \cdot r^n \Big) \\
= \;\; & \sum_{n=0}^{\infty} \sum_{m=-n}^{n} \left(\frac{(-1)^{n'} A_{n'}^0 \cdot A_n^m \cdot Y_{n+n'}^{-m}(\alpha,\beta)}{\rho^{n+n'+1} \cdot A_{n+n'}^m} \right) \cdot Y_n^m(\theta,\phi) r^n
\end{aligned}
$$

where the last equality is obtained from Lemma 3.3.1.

Case II : $m' < 0$. Using Lemmas 3.3.1 and 3.3.2, we have

$$\frac{Y_{n'}^{m'}(\theta', \phi')}{r'^{n'+1}}$$

$$= A_{n'}^{m'} \cdot \partial_{-P}^{|m'|} \partial_{z_P}^{n'-|m'|} \left(\frac{1}{r'} \right)$$

$$= \sum_{n=0}^{\infty} \Big(\sum_{m=-n}^{0} (-1)^{n'} A_{n'}^{m'} \cdot A_n^m \cdot \partial_{-Q}^{|m'|} \partial_{+Q}^{|m|} \partial_{z_Q}^{n+n'-|m|-|m'|} \left(\frac{1}{\rho} \right) \cdot Y_n^m(\theta, \phi) \cdot r^n +$$

$$\sum_{m=1}^{n} (-1)^{n'} A_{n'}^{m'} \cdot A_n^m \cdot \partial_{-Q}^{|m'|} \partial_{-Q}^{m} \partial_{z_Q}^{n+n'-|m'|-m} \left(\frac{1}{\rho} \right) \cdot Y_n^m(\theta, \phi) \cdot r^n \Big)$$

$$= \sum_{n=0}^{\infty} \sum_{m=-n}^{n} \left(\frac{J_m^{m'} \cdot A_{n'}^{m'} \cdot A_n^m \cdot Y_{n+n'}^{m'-m}(\alpha, \beta)}{\rho^{n+n'+1} \cdot A_{n+n'}^{m-m'}} \right) \cdot Y_n^m(\theta, \phi) \cdot r^n \; ,$$

where

$$J_m^{m'} = \begin{cases} (-1)^{n'}, & \text{if } m > 0; \\ (-1)^{n'}(-1)^{min(|m'|,|m|)}, & \text{if } m < 0. \end{cases} \tag{3.51}$$

Case III : $m' > 0$. From Lemma 3.3.1,

$$\frac{Y_{n'}^{m'}(\theta', \phi')}{r'^{n'+1}}$$

$$= A_{n'}^{m'} \cdot \partial_{+P}^{m'} \partial_{z_P}^{n'-m'} \left(\frac{1}{r'} \right)$$

$$= \sum_{n=0}^{\infty} \Big(\sum_{m=-n}^{0} (-1)^{n'} A_{n'}^{m'} \cdot A_n^m \cdot \partial_{+Q}^{m'} \partial_{+Q}^{-m} \partial_{z_Q}^{n+n'-(m'-m)} \left(\frac{1}{\rho} \right) \cdot Y_n^m(\theta, \phi) \cdot r^n +$$

$$\sum_{m=1}^{n} (-1)^{n'} A_{n'}^{m'} \cdot A_n^m \cdot \partial_{+Q}^{m'} \partial_{-Q}^{m} \partial_{z_Q}^{n+n'-m-m'} \left(\frac{1}{\rho} \right) \cdot Y_n^m(\theta, \phi) \cdot r^n \Big)$$

$$= \sum_{n=0}^{\infty} \sum_{m=-n}^{n} \left(\frac{J_m^{m'} \cdot A_{n'}^{m'} \cdot A_n^m \cdot Y_{n+n'}^{m'-m}(\alpha, \beta)}{\rho^{n+n'+1} \cdot A_{n+n'}^{m-m'}} \right) \cdot Y_n^m(\theta, \phi) \cdot r^n \; ,$$

where

$$J_m^{m'} = \begin{cases} (-1)^{n'}, & \text{if } m < 0; \\ (-1)^{n'}(-1)^{min(|m'|,|m|)}, & \text{if } m > 0. \end{cases} \tag{3.52}$$

As before, for the terms with $m > 0$, we have used Lemma 3.3.2 to annihilate whichever of the operators ∂_- and ∂_+ occurs less frequently.

The last addition theorem describes a formula for expanding a spherical harmonic of nonnegative degree about a shifted origin. Its proof is similar to those of the first two addition theorems. A more involved proof, based on group representation theory, can be found in [38].

THEOREM 3.5.3 (**Third Addition Theorem**) *Let $Q = (\rho, \alpha, \beta)$ be the center of expansion of an arbitrary spherical harmonic of nonnegative degree. Let the point $P = (r, \theta, \phi)$ with $P - Q = (r', \theta', \phi')$. Then*

$$Y_{n'}^{m'}(\theta', \phi') \cdot r'^{\,n'} = \sum_{n=0}^{n'} \sum_{m=-n}^{n} \frac{J_{n,m}^{m'} \cdot A_n^m \cdot A_{n'-n}^{m'-m} \cdot \rho^n \cdot Y_n^m(\alpha, \beta)}{A_{n'}^{m'}} \cdot Y_{n'-n}^{m'-m}(\theta, \phi) r^{n'-n}$$

$$(3.53)$$

where

$$J_{n,m}^{m'} = \begin{cases} (-1)^n (-1)^m, & \text{if } m \cdot m' < 0; \\ (-1)^n (-1)^{m'-m}, & \text{if } m \cdot m' > 0 \text{ and } |m'| < |m|; \\ (-1)^n, & \text{otherwise.} \end{cases}$$

$$(3.54)$$

We are in a position now to develop translation operators for spherical harmonic expansions.

THEOREM 3.5.4 (**Translation of a Multipole Expansion**) *Suppose that l charges of strengths $q_1, q_2, \ldots, q_l$ are located inside the sphere D of radius a with center at $Q = (\rho, \alpha, \beta)$, and that for points $P = (r, \theta, \phi)$ outside D, the potential due to these charges is given by the multipole expansion*

$$\Phi(P) = \sum_{n=0}^{\infty} \sum_{m=-n}^{n} \frac{O_n^m}{r'^{\,n+1}} \cdot Y_n^m(\theta', \phi') ,$$

$$(3.55)$$

where $P - Q = (r', \theta', \phi')$. Then for any point $P = (r, \theta, \phi)$ outside the sphere D_1 of radius $(a + \rho)$,

$$\Phi(P) = \sum_{j=0}^{\infty} \sum_{k=-j}^{j} \frac{M_j^k}{r^{j+1}} \cdot Y_j^k(\theta, \phi)$$

$$(3.56)$$

where

$$M_j^k = \sum_{n=0}^{j} \sum_{m=-n}^{n} \frac{O_{j-n}^{k-m} \cdot J_m^{k-m} \cdot A_n^m \cdot A_{j-n}^{k-m} \cdot \rho^n \cdot Y_n^{-m}(\alpha, \beta)}{A_j^k} ,$$

$$(3.57)$$

with J_r^s and A_r^s defined by equations (3.43) and (3.26), respectively. Furthermore, for any $p \geq 1$,

$$\left| \Phi(P) - \sum_{j=0}^{p} \sum_{k=-j}^{j} \frac{M_j^k}{r^{j+1}} \cdot Y_j^k(\theta, \phi) \right| \leq \left(\frac{\sum_{i=1}^{l} |q_i|}{r - (a + \rho)} \right) \left(\frac{a + \rho}{r} \right)^{p+1} \qquad (3.58)$$

Proof : The coefficients of the shifted expansion (3.56) are obtained by applying the First Addition Theorem to each of the terms in the original expansion (3.55). For the error bound (3.58), observe that the terms M_n^m are the coefficients of the (unique) multipole expansion about the origin of those charges contained in the sphere D, and Theorem 3.4.2 applies with a replaced by $a + \rho$.

The second translation procedure is used to convert a multipole expansion of the field induced by a collection of charges into a local expansion inside some region of analyticity.

THEOREM 3.5.5 (Conversion of a Multipole Expansion into a Local Expansion) *Suppose that l charges of strengths $q_1, q_2, \cdots, q_l$ are located inside the sphere D_Q of radius a with center at $Q = (\rho, \alpha, \beta)$, and that $\rho > (c+1)a$ with $c > 1$ (Figure 3.3). Then the corresponding multipole expansion (3.55) converges inside the sphere D_0 of radius a centered at the origin. Inside D_0, the potential due to the charges $q_1, q_2, \cdots, q_l$ is described by a local expansion:*

$$\Phi(P) = \sum_{j=0}^{\infty} \sum_{k=-j}^{j} L_j^k \cdot Y_j^k(\theta, \phi) \cdot r^j , \qquad (3.59)$$

where

$$L_j^k = \sum_{n=0}^{\infty} \sum_{m=-n}^{n} \frac{O_n^m \cdot J_k^m \cdot A_n^m \cdot A_j^k \cdot Y_{j+n}^{m-k}(\alpha, \beta)}{A_{j+n}^{m-k} \cdot \rho^{j+n+1}} , \qquad (3.60)$$

with J_r^s and A_r^s defined by equations (3.49) and (3.26), respectively. Furthermore, for any $p \geq 1$,

$$\left| \Phi(P) - \sum_{j=0}^{p} \sum_{k=-j}^{j} L_j^k \cdot Y_j^k(\theta, \phi) \cdot r^{j+1} \right| \leq \left(\frac{\sum_{i=1}^{l} |q_i|}{ca - a} \right) \left(\frac{1}{c} \right)^{p+1} . \qquad (3.61)$$

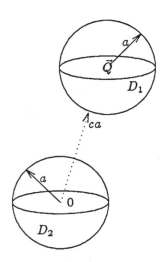

Figure 3.3
Source charges $q_1, q_2, ..., q_l$ are contained in the sphere D_1. The corresponding
multipole expansion about Q converges inside D_2.

Proof :   We obtain the coefficients of the local expansion (3.59) by applying the
Second Addition Theorem to each of the terms in the multipole expansion (3.55).
The bound (3.61) is an immediate consequence of the simpler error bound (3.16)
and the triangle inequality.

The following theorem yields a procedure for shifting the origin of a truncated
local expansion. It is an exact translation, and no error bound is needed.

THEOREM 3.5.6(**Translation of a Local Expansion**)
Let $Q = (\rho, \alpha, \beta)$ be the origin of a local expansion

$$\Phi(P) = \sum_{n=0}^{p} \sum_{m=-n}^{n} O_n^m \cdot Y_n^m(\theta', \phi') \cdot r'^n \, , \tag{3.62}$$

where $P = (r, \theta, \phi)$ and $P - Q = (r', \theta', \phi')$. Then

$$\Phi(P) = \sum_{j=0}^{p} \sum_{k=-j}^{j} L_j^k \cdot Y_j^k(\theta, \phi) \cdot r^j \, , \tag{3.63}$$

where

$$L_j^k = \sum_{n=j}^{p} \sum_{m=-n}^{n} \frac{O_n^m \cdot J_{n-j,m-k}^m \cdot A_{n-j}^{m-k} \cdot A_j^k \cdot Y_{n-j}^{m-k}(\alpha,\beta) \cdot \rho^{n-j}}{A_n^m} \, , \qquad (3.64)$$

with $J_{r,s}^t$ and A_r^s defined by equations (3.54) and (3.26), respectively.

Proof : The coefficients (3.64) are obtained by applying the Third Addition Theorem to each of the terms in the expansion (3.62).

3.6 The Fast Multipole Algorithm

We describe here the analog of the non-adaptive algorithm of Chapter 2. The computational box is depicted in Figure 3.4. It is a cube with sides of length one, centered about the origin of the coordinate system, and is assumed to contain all N particles of the system under consideration.

Fixing a precision ϵ, we choose $p = \lceil -\log_2(\epsilon) \rceil$ and specify that no interactions be computed by means of multipole expansions for clusters of particles which are not contained in well-separated spheres. This is precisely the condition needed for the error bounds (3.38),(3.58) and (3.61) to apply with $c = 2$, the truncation error to be bounded by 2^{-p}, and the desired precision to be achieved. In order to impose such a condition, we introduce a hierarchy of meshes which refine the computational box into smaller and smaller regions (Figure 3.4). Mesh level 0 is equivalent to the entire box, while mesh level $l + 1$ is obtained from level l by subdivision of each region into eight equal parts. The number of distinct boxes at mesh level l is equal to 8^l. A tree structure is imposed on this mesh hierarchy, so that if *ibox* is a fixed box at level l, the eight boxes at level $l + 1$ obtained by subdivision of *ibox* are considered its children.

Other notation used in the description of the algorithm includes

nearest neighbor: For box i at level l, a nearest neighbor is a box at the same level of refinement which shares a boundary point with box i.

second nearest neighbor: For box i at level l, a second nearest neighbor is a box at the same level of refinement which shares a boundary point with a nearest neighbor of box i.

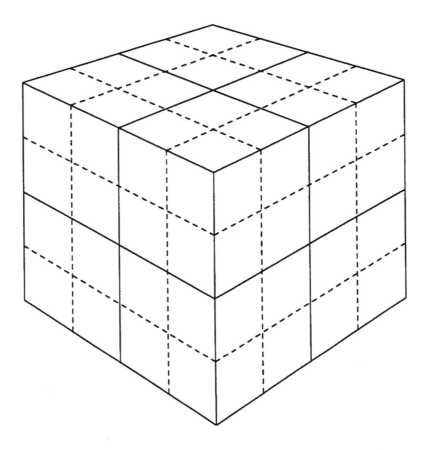

Figure 3.4
The computational box and the first two levels of refinement, indicated by the solid and dashed lines, respectively.

$\Phi_{l,i}$: the pth-order multipole expansion (about the box center) of the potential field created by the particles contained inside box i at level l,

$\Psi_{l,i}$: the pth-order local expansion about the center of box i at level l, describing the potential field due to all particles outside the box and its nearest and second nearest neighbors,

$\tilde{\Psi}_{l,i}$: the pth-order local expansion about the center of box i at level l, describing the potential field due to all particles outside i's *parent* box and the *parent* box's nearest and second nearest neighbors.

Interaction list: for box i at level l, it is the set of boxes which are children of the nearest and second nearest neighbors of i's parent and which are not nearest or second nearest neighbors of box i.

We first observe that if boxes A and B are in each other's interaction lists, then the smallest spheres S_A and S_B containing these boxes are well-separated in the sense described in section 3.4, so the desired error bounds apply. Suppose now that at level $l-1$, the local expansion $\Psi_{l-1,i}$ has been obtained for all boxes. Then, by using Theorem 3.5.4 to shift (for all i) the expansion $\Psi_{l-1,i}$ to each of box i's children, we have, for each box j at level l, a local representation of the potential due to all particles outside of j's parent's nearest and second nearest neighbors, namely $\tilde{\Psi}_{l,j}$. The interaction list is, therefore, precisely that set of boxes whose contribution to the potential must be added to $\tilde{\Psi}_{l,j}$ in order to create $\Psi_{l,j}$. This is done by using Theorem 3.5.5 to convert the multipole expansions of these interaction boxes to local expansions about the current box center and adding them to the expansion obtained from the parent. Note also that with free-space boundary conditions, $\Psi_{0,i}$ and $\Psi_{1,i}$ are equal to zero since there are no well-separated boxes to consider, and we can begin forming local expansions at level 2.

Following is a formal description of the algorithm.

Algorithm

Initialization

Choose a level of refinement $n \approx \log_8 N$, a precision ϵ, and set $p = \lceil -\log_2(\epsilon) \rceil$.

Upward Pass

Step 1

Comment [Form multipole expansions of potential field due to particles
in each box about the box center at the finest mesh level.]

do $ibox = 1, ..., 8^n$
    Form a pth-degree multipole expansion $\Phi_{n,ibox}$, by using Theorem 3.4.2.
enddo

Step 2

Comment [Form multipole expansions about the centers of all boxes
at all coarser mesh levels, each expansion representing the potential
field due to all particles contained in one box.]

do $l = n - 1, ..., 0$
    do $ibox = 1, ..., 8^l$
        Form a pth-degree multipole expansion $\Phi_{l,ibox}$, by using
        Theorem 3.5.4 to shift the center of each child box's expansion
        to the current box center and adding them together.
    enddo
enddo

Downward Pass

Comment [In the downward pass, interactions are consistently computed
at the coarsest possible level. For a given box, this is accomplished
by including interactions with those boxes which are well-separated
and whose interactions have not been accounted for at the parent's
level.]

Step 3

Comment [Form a local expansion about the center of each box at each mesh
level $l \le n - 1$. This local expansion describes the field due to all
particles in the system that are not contained in the current box, its
nearest neighbors, or its second nearest neighbors. Once the local
expansion is obtained for a given box, it is shifted, in the second
inner loop to the centers of the box's children, forming the initial
expansion for the boxes at the next level.]

Set $\tilde{\Psi}_{1,1} = \tilde{\Psi}_{1,2} = \cdots = \tilde{\Psi}_{1,8} = (0,0,...,0)$
do $l = 1, ..., n - 1$
    do $ibox = 1, ..., 8^l$
        Form $\Psi_{l,ibox}$ by using Theorem 3.5.5 to convert the multipole
        expansion $\Phi_{l,j}$ of each box j in *interaction list* of box *ibox*
        to a local expansion about the center of box *ibox*, adding these
        local expansions together, and adding the result to $\tilde{\Psi}_{l,ibox}$.
    enddo

do $ibox = 1, ..., 8^l$
   Form the expansion $\tilde{\Psi}_{l+1,j}$ for $ibox$'s children
   by using Theorem 3.5.6 to expand $\Psi_{l,ibox}$ about the children's box centers.
   enddo
enddo

Step 4

Comment [Compute interactions at finest mesh level]

do $ibox = 1, ..., 8^n$
   Form $\Psi_{n,ibox}$ by using Theorem 3.5.5 to convert the multipole
   expansion $\Phi_{n,j}$ of each box j in *interaction list* of box $ibox$
   to a local expansion about the center of box $ibox$, adding these
   local expansions together, and adding the result to $\tilde{\Psi}_{n,ibox}$.
enddo

Comment [Local expansions at finest mesh level are now available.
   They can be used to generate the potential or force due to all
   particles outside the nearest and second nearest neighbor boxes
   at the finest mesh level.]

Step 5

Comment [Evaluate local expansions at particle positions.]

do $ibox = 1, ..., 8^n$
   For every particle p_j located at the point P_j in box $ibox$,
   evaluate $\Psi_{n,ibox}(P_j)$.
enddo

Step 6

Comment [Compute potential (or force) due to near neighbors directly.]

do $ibox = 1, ..., 8^n$
   For every particle p_j in box $ibox$, compute interactions with
   all other particles within the box and its nearest and second
   nearest neighbors.
enddo

Step 7

do $ibox = 1, ..., 8^n$
   For every particle in box $ibox$, add direct and far-field terms together.
enddo

Remark: Each local expansion is described by its p^2 coefficients. Direct evaluation of this expansion at a point yields the potential. But the force can be obtained from the gradient of the local expansion, and these partial derivatives are available analytically. There is no need for numerical differentiation. Furthermore, since the components of $\nabla\Phi$ are themselves harmonic, there exist error bounds for the force of exactly the same form as (3.38),(3.58) and (3.61).

A brief analysis of the algorithmic complexity is given below.

Step Number	Operation Count	Explanation
Step 1	order Np^2	each particle contributes to one expansion at the finest level.
Step 1	order Np^2	each particle contributes to one expansion at the finest level.
Step 2	order Np^4	At the lth level, 8^l shifts involving order p^4 work per shift must be performed.
Step 3	order $\leq 876Np^4$	There are at most 875 entries in the interaction list for each box at each level. An extra order Np^4 work is required for the second loop.
Step 4	order $\leq 875Np^4$	Again, there are at most 875 entries in the interaction list for each box and $\approx N$ boxes.
Step 5	order Np^2	One p^{th}-degree expansion is evaluated for each particle.
Step 6	order $\frac{25}{2}Nk_n$	Let k_n be a bound on the number of particles per box at the finest mesh level. Interactions must be computed within the box

and its eight nearest neighbors,
but using Newton's third law,
we need only compute half
of the pairwise interactions.

Step **7**   order N Adding two terms for each particle.

The estimate for the running time is therefore

$$N \cdot (a \cdot p^2 + b \cdot p^4 + d \cdot k_n + e) \, ,$$

with the constants $a, b, c, d,$ and e determined by the computer system, language, implementation, etc.

As in the two-dimensional case, the asymptotic storage requirements of the algorithm are largely dependent on the number of boxes created. In particular, the pth-degree expansions $\Phi_{l,j}$ and $\Psi_{l,j}$ must be stored for every box at every level. We must also store the locations of the particles, their charges, and the results of the calculations (the potentials and/or electric fields). The net storage requirements are therefore of the form

$$(\alpha + \beta \cdot p^2) \cdot N \, ,$$

with the coefficients α and β determined, as above, by the computer system, language, implementation, etc.

4 Numerical results

A computer program in Fortran 77 has been implemented using the two-dimensional algorithms of the second chapter. It is capable of handling free-space problems, and problems with periodic, homogeneous Dirichlet, or homogeneous Neumann boundary conditions. All calculations cited below have been carried out on a VAX-8600, running VMS version 4.3 .

In the first set of experiments, we considered free-space problems with a variety of particle distributions. For each distribution, the corresponding fields were computed in four ways: by the adaptive algorithm in single precision, by the homogeneous algorithm in single precision, and directly in single and double precision. The direct calculation of the field in double precision was used as a standard for comparing the relative accuracies of the other three methods. In these experiments, the number of particles varied between 100 and 25600, with charge strengths randomly assigned between zero and one.

The results are summarized in Tables 4.1, 4.2, 4.3, and 4.4. The first column of each table contains the number of particles N for which calculations have been performed. In the remaining columns, the upper case letters T, E and S are used to denote the corresponding computational time, error and storage, with the subscripts *alg, uni* and *dir* referring to the adaptive algorithm, the non-adaptive algorithm, and the direct (single-precision) calculation respectively. More specifically, columns 2 through 4 show the times, in seconds, required to compute the field by the three methods. The errors E_{alg}, E_{uni} and E_{dir} for the adaptive, non-adaptive and direct method, respectively, are presented in the next three columns. They are defined by the formula

$$E = \left(\frac{\sum_{i=1}^{N} \left| f_i - \tilde{f}_i \right|^2}{\sum_{i=1}^{N} |f_i|^2} \right)^{1/2}$$

where f_i is the value of the field at the i-th particle position obtained by direct calculation in double precision and $\tilde{f}_i$ is the result obtained by one of the three methods being studied. The last two columns of the tables contain the storage requirements S_{alg} and S_{uni}, in single-precision words, for the two fast multipole methods.

Remark: For the tests involving 12800 and 25600 particles, it was not considered practical to use the direct method to calculate the fields at all particle positions, since this would require prohibitive amounts of CPU time without providing much useful information. Therefore, we have performed the direct calculations in double precision for only 100 of the particles, and used these results to evaluate the relative accuracies. The corresponding values of T_{dir} were estimated by extrapolation.

For the first set of tests, the positions of the charged particles were randomly distributed in a square, and the resulting particle density was roughly uniform (Figure 4.1). The number of terms in the expansions was set to 20, and the maximum number of particles in a childless box was set to 30.

In the second set of experiments, the charged particles were distributed along a curve (Figure 4.2). The number of terms in the expansions was set to 17 and the maximum number of particles in a childless box was set to 30.

The third set of numerical experiments was performed on extremely non-uniform distributions of particles (Figure 4.3). A fifth of the N particles were randomly assigned in a square of area one. Two fifths were randomly distributed about the center of the square in a circle of radius 0.003. The rest of the particles were assigned positions inside a circle of radius 0.5 with a density inversely proportional to the square of the distance from the center. The number of terms in the expansions was set to 17 and the maximum number of particles per childless box was set to 30.

In the last set of free-space experiments, half of the particles were distributed along a curve similar to that of the second set of experiments and the rest of the particles were distributed inside four circles with a density inversly proportional to the square of the distance from the centers of the circles (Figure 4.4). The number of terms in the expansions was set to 17 and the maximum number of particles per childless box was set to 30.

The following observations can be made from Tables 4.1, 4.2, 4.3 and 4.4, where the results of the experiments described above are summarized.

1. The accuracies of the results obtained by the algorithms using multipole expansions are in agreement with the error bounds given in (2.10),(2.21) and (2.25). For the most part, the fast methods are slightly more accurate than the direct calculation.

2. In all cases, the actual CPU time requirements of the adaptive algorithm grow linearly with N. The CPU time requirements of the non-adaptive scheme grow linearly for homogeneous distributions, but not for extremely non-uniform distributions (see Tables 4.3 - 4.4).

3. Even for uniform distributions of charges, the adaptive algorithm is about 30% faster than the non-adaptive one.

4. The storage requirements of both fast algorithms are roughly proportional to the number of particles involved in the simulations. The storage requirements of the adaptive algorithm are about four times less than those of the non-adaptive version.

5. By the time the number of particles reaches 25600, the adaptive algorithm is about 100 times faster than the direct method for the case of a uniform distribution (see Table 4.1). When the charges are situated on a curve, the adaptive scheme is roughly 200 times faster than the direct method, and about 3 times faster than the non-adaptive scheme(see Table 4.2).

6. For the highly non-uniform case (see Table 4.3), the adaptive algorithm is slightly more efficient than for the uniform distribution. The non-adaptive scheme displays an almost quadratic growth of CPU time with N, and is about 25 times slower than its adaptive counterpart by the time $N = 25600$.

7. Even for as few as 1600 particles, the adaptive algorithm is about ten times faster than the direct calculation.

8. The performance of the algorithm does not depend on the shape of the region where the charges are distributed (see Table 4.4.)

Similar calculations have been performed for periodic, homogeneous Dirichlet and Neumann boundary conditions, and the observations made above are equally applicable in these cases.

For illustration, the equipotential lines for a box with 10 randomly distributed particles and Dirichlet boundary conditions are shown in Figure 4.5. The entire calculation required 15 seconds of CPU time; about half the time was spent evaluating the field at more than 10,000 points, while the rest was used up by the plotting routine.

Figure 4.1
25600 uniformly located charges in the computational cell.

N	T_{alg}	T_{uni}	T_{dir}	E_{alg}	E_{uni}	E_{dir}	S_{alg}	S_{uni}
100	0.15	0.47	0.15	$1.7\ 10^{-6}$	$4.0\ 10^{-7}$	$1.7\ 10^{-6}$	866	4179
200	0.43	0.65	0.61	$9.3\ 10^{-7}$	$4.3\ 10^{-7}$	$4.4\ 10^{-7}$	2503	5479
400	1.01	1.94	2.47	$7.0\ 10^{-7}$	$6.4\ 10^{-7}$	$6.4\ 10^{-7}$	3763	16847
800	2.45	2.78	10.27	$4.1\ 10^{-7}$	$4.0\ 10^{-7}$	$4.7\ 10^{-7}$	11203	22047
1600	5.37	8.56	42.35	$3.7\ 10^{-7}$	$4.2\ 10^{-7}$	$5.4\ 10^{-7}$	15923	67519
3200	10.60	11.80	152.95	$5.0\ 10^{-7}$	$5.3\ 10^{-7}$	$8.7\ 10^{-7}$	44423	88319
6400	23.38	33.49	601.18	$7.0\ 10^{-7}$	$5.4\ 10^{-7}$	$1.3\ 10^{-6}$	65907	270207
12800	45.34	48.02	2433.20	$6.0\ 10^{-7}$	$4.9\ 10^{-7}$	$1.6\ 10^{-6}$	176631	353407
25600	96.72	137.68	9694.45	$8.3\ 10^{-7}$	$8.9\ 10^{-7}$	$2.2\ 10^{-6}$	268723	1080959

Table 4.1
Uniformly distributed particles. $p = 20$ and $s = 30$.

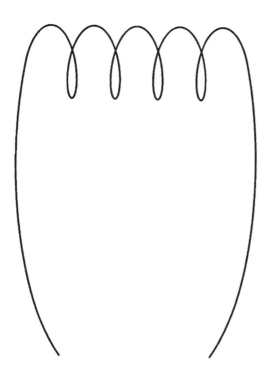

Figure 4.2
6400 particles distributed on a curve.

N	T_{alg}	T_{uni}	T_{dir}	E_{alg}	E_{uni}	E_{dir}	S_{alg}	S_{uni}
100	0.11	0.38	0.16	$3.4\ 10^{-5}$	$3.2\ 10^{-5}$	$3.4\ 10^{-5}$	1149	3927
200	0.30	0.54	0.57	$8.9\ 10^{-6}$	$9.3\ 10^{-6}$	$8.9\ 10^{-6}$	2694	5227
400	0.64	1.31	2.29	$5.6\ 10^{-5}$	$5.6\ 10^{-5}$	$5.6\ 10^{-5}$	5103	15827
800	1.46	3.13	9.30	$9.4\ 10^{-5}$	$9.5\ 10^{-5}$	$9.5\ 10^{-5}$	10133	21027
1600	2.66	5.94	37.41	$2.0\ 10^{-5}$	$2.0\ 10^{-5}$	$2.0\ 10^{-5}$	19241	63427
3200	5.93	12.50	149.21	$7.8\ 10^{-6}$	$8.7\ 10^{-6}$	$8.8\ 10^{-6}$	40055	84227
6400	12.42	29.66	597.95	$4.2\ 10^{-5}$	$4.2\ 10^{-5}$	$4.2\ 10^{-5}$	84429	253827
12800	25.11	79.47	2425.48	$8.7\ 10^{-5}$	$8.7\ 10^{-5}$	$8.8\ 10^{-5}$	167421	337027
25600	47.53	152.07	9581.20	$8.9\ 10^{-5}$	$9.1\ 10^{-5}$	$8.9\ 10^{-5}$	332927	1015427

Table 4.2
Particles distributed on a curve. $p = 17$ and $s = 30$.

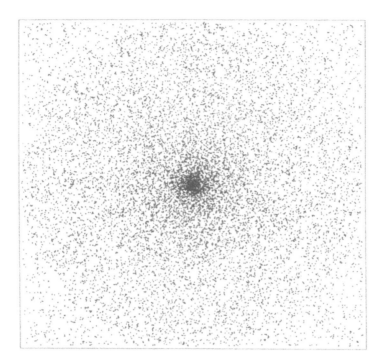

Figure 4.3
Highly non-uniform distribution of 25600 charges.

N	T_{alg}	T_{uni}	T_{dir}	E_{alg}	E_{uni}	E_{dir}	S_{alg}	S_{uni}
100	0.19	0.45	0.15	$2.7\,10^{-6}$	$1.0\,10^{-5}$	$2.8\,10^{-6}$	2508	3927
200	0.48	0.74	0.57	$6.9\,10^{-6}$	$7.6\,10^{-6}$	$6.9\,10^{-6}$	4014	5227
400	1.13	2.26	2.33	$1.9\,10^{-6}$	$9.0\,10^{-6}$	$1.9\,10^{-6}$	8307	15827
800	2.25	5.15	9.34	$4.3\,10^{-6}$	$6.0\,10^{-6}$	$3.7\,10^{-6}$	13353	21027
1600	5.09	16.17	37.74	$2.4\,10^{-6}$	$1.6\,10^{-6}$	$2.1\,10^{-6}$	25588	63427
3200	9.98	50.23	149.86	$3.7\,10^{-6}$	$1.4\,10^{-6}$	$1.7\,10^{-6}$	46806	84227
6400	21.80	177.13	606.14	$5.8\,10^{-6}$	$4.0\,10^{-6}$	$5.9\,10^{-6}$	90505	253827
12800	41.93	663.21	2420.33	$4.0\,10^{-6}$	$4.0\,10^{-6}$	$4.2\,10^{-6}$	186226	337027
25600	90.05	2317.93	9622.63	$2.9\,10^{-6}$	$3.0\,10^{-6}$	$4.0\,10^{-6}$	373639	1015427

Table 4.3
Highly non-uniform distribution of particles. $p = 17$ and $s = 30$.

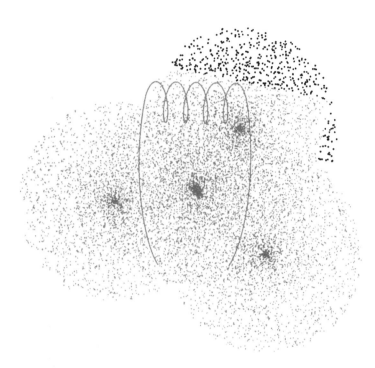

Figure 4.4
A non-uniform distribution of 25600 charges in a region of complicated shape.

N	T_{alg}	T_{uni}	T_{dir}	E_{alg}	E_{uni}	E_{dir}	S_{alg}	S_{uni}
100	0.15	0.43	0.15	$4.3\ 10^{-5}$	$5.5\ 10^{-5}$	$5.0\ 10^{-5}$	1145	3927
200	0.39	0.68	0.59	$3.3\ 10^{-5}$	$3.9\ 10^{-5}$	$3.3\ 10^{-5}$	3224	5227
400	0.84	1.69	2.31	$8.1\ 10^{-5}$	$7.1\ 10^{-6}$	$8.1\ 10^{-5}$	6939	15827
800	2.11	5.03	9.39	$4.3\ 10^{-5}$	$4.3\ 10^{-5}$	$4.3\ 10^{-5}$	13406	21027
1600	4.35	11.34	37.74	$9.2\ 10^{-5}$	$9.2\ 10^{-5}$	$9.2\ 10^{-5}$	24913	63427
3200	9.16	30.85	153.76	$1.1\ 10^{-5}$	$1.1\ 10^{-5}$	$1.1\ 10^{-5}$	48902	84227
6400	19.22	48.62	611.82	$5.4\ 10^{-6}$	$5.5\ 10^{-6}$	$5.4\ 10^{-6}$	96153	253827
12800	37.92	155.75	2440.90	$2.1\ 10^{-5}$	$2.0\ 10^{-5}$	$2.1\ 10^{-5}$	194377	337027
25600	80.02	248.90	9798.34	$4.4\ 10^{-5}$	$4.4\ 10^{-5}$	$4.5\ 10^{-5}$	388624	1015427

Table 4.4
Non-uniform distribution of particles in a region of complicated shape. $p = 17$ and $s = 30$.

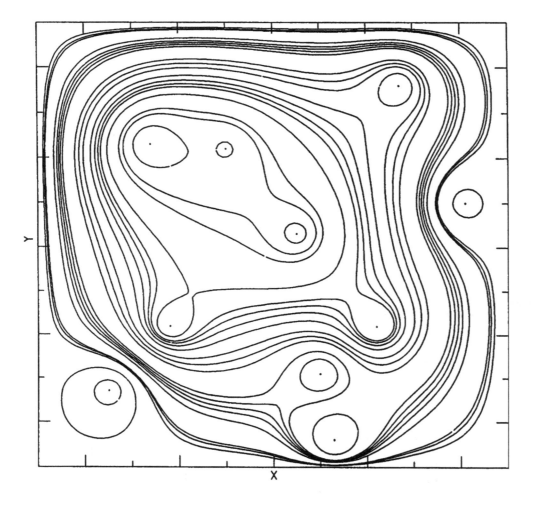

Figure 4.5
The equipotential lines for the electrostatic field due to 10 randomly located charges
in the computational cell, with homogeneous Dirichlet boundary conditions.

5 Conclusions and Applications

We have constructed several algorithms for the rapid evaluation of the potential and force fields generated by systems of particles whose interactions are Coulombic or gravitational in nature. These algorithms are applicable in both two and three dimensions, and allow for the solution of free-space problems, as well as problems with periodic, homogeneous Dirichlet, and homogeneous Neumann boundary conditions. The asymptotic CPU time estimate for these algorithms is of the order $O(N)$, where N is the number of particles in the system, and the numerical examples we present in Chapter 4 indicate that even very large-scale problems result in acceptable CPU time requirements.

The study of a number of physical systems has benefited from particle simulation. Below, we list several areas where the approach described in this dissertation offers advantages over previously published methods.

Astrophysics

Several interesting questions in cosmology have been investigated by the computer simulation of many-body systems whose interactions are governed by Newton's law of gravitation. Early work on the evolution of stellar clusters by Aarseth and others used the direct method to compute the necessary N^2 interactions at every time step [1]. More recent work using particle/mesh methods [24] or clustering methods [7] allowed simulations with an order of magnitude more bodies. Both methods achieve an increase in computational speed at the expense of accuracy.

Plasma Physics

In particle models for plasma simulation, one follows the motion of a large number of charged particles in their self-consistent electric and magnetic fields [13,24]. Simulations using particle/mesh methods have greatly enhanced our understanding of collective phenomena in plasmas. Unfortunately, there are several interesting problems which have been essentially unapproachable with these methods, due to the uniformity of the grid, aliasing, and smoothing. These include the simulation of "cold plasmas" and ion beams. In addition, free space and exterior problems have been difficult to handle, since the grids require the imposition of a boundary whether it is desired or not.

Molecular Dynamics

Molecular dynamics is a technique for studying the properties of fluids (and other phases of matter) by computer simulation. Once initial positions and velocities are chosen for some number of representative particles, their trajectories are followed by integrating Newton's second law of motion. In general, the particles are chosen to be individual atoms or molecules which constitute the material under consideration. Much has been learned from molecular dynamics about the fine structure and thermodynamics of water, aqueous solutions, and a variety of other polar and non-polar liquids [2,3,6,34]. Recent work has extended the method to more complicated systems, such as polymers in solution [9], lipid bilayers [31], and proteins [26,28].

In early simulations, only non-polar fluids were considered, with either "hard-sphere" particle-particle interactions [4], or interactions governed by a Lennard-Jones potential [29,33]

$$\Phi(r_{ij}) = 4\epsilon \left[\left(\frac{\sigma}{r_{ij}} \right)^{12} - \left(\frac{\sigma}{r_{ij}} \right)^{6} \right] . \qquad (5.1)$$

At long range, this interaction has an attraction proportional to r^{-6}, while at close range, there is a r^{-12} repulsion. Because of the rapid decay, in most simulations, interactions are accounted for up to a fixed truncation radius. As a result, the amount of computation per time step is proportional to the number of particles N.

In polar fluids, the situation is quite different. A Coulombic term is added to the potential function Φ in (5.1), and all pairwise interactions should be accounted for. If periodic boundary conditions are used, it is also necessary to include the effects of all image charges.

The usual approach taken to increasing the allowable number of particles in simulations of polar fluids has simply been to truncate the potential at some fixed cut-off distance on empirical grounds. There are many papers in the literature using this approximation [6,8,34]. Little error seems to be introduced in the (local) atom-atom correlation functions, but dielectric properties are poorly simulated [2,30]. The answer to the question of why these dielectric properties are incorrectly computed is currently unknown. The difficulty may be in truncation, it may be that the number of particles has not reached the asymptotic range, and it may be that the semi-classical molecular dynamics model is insufficient. The use of the fast multipole method should help to distinguish between these cases since it allows for rapid large-scale calculations without any error due to truncation.

Fluid Dynamics

The governing equations for viscous fluid flow are the Navier-Stokes equations. The vortex blob method of Chorin [12] is a grid-free numerical method for the solution of these equations. For a detailed description, we refer the reader to the original work. We merely indicate here that the dominant cost per time step is the computation of the interactions between N vortex blobs. Over large distances, these interactions are Coulombic (simulating point vortices), while at close range the interactions take a different form. Recalling that in the fast multipole method we compute nearest neighbor interactions directly, there is no difficulty in changing the local behavior of the field. The fast multipole method therefore provides a significant reduction in what has been the dominant computational cost of the vortex method.

Elliptic Partial Differential Equations

Boundary value problems for the Laplace equation can be reduced to integral equations of the second kind by means of classical potential theory [25]. For example, to solve the Dirichlet problem

$$\nabla^2 \Phi(p) = 0 \quad \text{in} \quad \Omega$$

$$\Phi(p) = f(p) \quad \text{on} \quad \partial\Omega \,,$$

we try to determine a function $\sigma(t)$ such that

$$\Phi(p) = \int_{\partial\Omega} \frac{\partial}{\partial n} G(p,t) \cdot \sigma(t) \; dt \,,$$

where $G(p,t)$ is the potential at p due to a unit charge at t. To satisfy the boundary condition, we must have

$$\sigma(p) + \int_{\partial\Omega} \frac{\partial}{\partial n} G(p,t) \cdot \sigma(t) \; dt = f(p) \tag{5.2}$$

(see, for example, [25]). Discretization of this integral equation leads to a large scale system of linear algebraic equations, which are in turn solved by some iterative technique. Most iterative schemes for the solution of linear systems resulting from classical potential theory require the application of the matrix to a recursively generated sequence of vectors. Applying a dense matrix to a vector requires order $O(n^2)$ operations, where n is the order of the system.

In this case, the dimension of the system is equal to the number of nodes in the discretization of the boundary $\partial\Omega$. As a result, the whole process is at least of the order $O(n^2)$. However, the matrix-vector product corresponds to evaluation of the field due to n sources located on the boundary, at each of the source locations themselves. It is therefore possible, by using the fast multipole method, to solve the original equation (5.2) in an amount of time proportional to n. We also note that once the integral equation has been solved and the source density $\sigma(t)$ determined, the field Φ can be evaluated at m interior points in order $O(m + n)$ operations. The rapid solution of integral equations by this approach was originally reported by Rokhlin [36].

Let us now consider more general elliptic partial differential equations of the form

$$\nabla^2\Phi + \alpha \cdot \frac{\partial\Phi}{\partial x} + \beta \cdot \frac{\partial\Phi}{\partial y} + \gamma \cdot \Phi = \delta \ ,$$

where α, β, γ, and δ are functions of position. The solution can be represented in the form

$$\Phi(p) = \int_{\partial\Omega} H(p, t) \cdot \sigma(t) \ dt + \int_{\Omega} H(p, \tau) \cdot \rho(\tau) \ d\tau$$

(see, for example, [37]). In order to satisfy the differential equation and whatever boundary condition has been imposed, a system of two integral equations for σ and ρ must be solved. Since the function $H(p, t)$ is taken to be the field at p due to a unit charge located at t (or some higher derivative), it is clear that the fast multipole method allows for the rapid solution of a broad class of such problems.

Numerical Complex Analysis

Many problems in complex analysis can be reduced to that of computing a Cauchy integral

$$f(z) = \frac{1}{2\pi i} \int_{\Gamma} \frac{h(t)}{t - z} dt \tag{5.3}$$

where Γ denotes the boundary of some domain D in the complex plane. Examples include the evaluation of analytic functions, the solution of boundary value problems for harmonic functions, and conformal mapping [22,40].

Discretizing the boundary curve into N equal segments, we have

$$f(z) \approx \frac{\Delta t}{2\pi i N} \sum_{j=1}^{N} \frac{h(t_j)}{t_j - z} \tag{5.4}$$

In the situations mentioned above, we generally need to evaluate this function f at some number of distinct points z_k, $k = 1, \ldots, M$. But $h(t_j)/(t_j - z_k)$ is just the electrostatic field at z_k due to a charge of strength $h(t_j)$ located at t_j. By using the fast multipole method, it is therefore possible to compute f at the M points z_k at a cost proportional to $(M + N)$, rather than $M \cdot N$.

When $M = N$, and $z_k = t_k$, this calculation can be viewed as the application of a Hilbert matrix to a vector. Given a collection of points $z_1, \ldots, z_n$ in $\mathbb{C}$, the Hilbert matrix associated with the points $\{z_i\}$ is defined as follows:

$$A_{ij} = \frac{1}{z_i - z_j} \quad \text{for} \quad i \neq j,$$

$$A_{ii} = 0.$$

The fast multipole method therefore provides an order $O(N)$ procedure for applying an $N \times N$ Hilbert matrix to an arbitrary vector. The question of whether it is possible to compute this matrix-vector product in fewer than $O(N^2)$ operations has recently been posed as the Trummer problem [16,17, 18,35].

Bibliography

[1] S. J. Aarseth, *Dynamical Evolution of Clusters of Galaxies - I*, Mon. Not. R. Astron. Soc. 126 (1963), pp. 223-255 .

[2] D. J. Adams and E. M. Adams, *Static dielectric properties of the Stockmayer fluid from Computer Simulation*, Molecular Phys., 42 (1981), pp. 907-926 .

[3] D. J. Adams, E. M. Adams and G. J. Hills, *The Computer Simulation of Polar Liquids*, Molec. Phys., 38 (1979), pp. 387-400.

[4] B. J. Alder and T. E. Wainwright, *Studies in Molecular Dynamics. I. General Method*, J. Chem. Phys., 31 (1959), pp. 459-466.

[5] C. R. Anderson, *A Method of Local Corrections for Computing the Velocity Field Due to a Distribution of Vortex Blobs*, J. Comput. Phys., 62 (1986), pp. 111-123.

[6] T. A. Andrea, W. C. Swope and H. C. Andersen, *The Role of Long Ranged Forces in Determining the Structure and Properties of Liquid Water*, J. Chem. Phys., 79 (1983), pp. 4576-4584.

[7] A. W. Appel, *An Efficient Program for Many-body Simulation*, Siam. J. Sci. Stat. Comput., 6 (1985), pp. 85-103.

[8] B. R. Brooks, R. E. Bruccoleri, B. D. Olafson, D. J. States, S. Swaminathan, and M. Karplus, *CHARMM: A Program for Macromolecular Energy, Minimization, and Dynamics Calculations*, J Comput. Chem., 4 (1983), pp. 187-217.

[9] W. Bruns and R. Bansal, *Molecular Dynamics Study of a Single Polymer Chain in Solution*, J Chem. Phys., 74 (1981), pp. 2064-2072.

[10] B. C. Carlson and G. S. Rushbrooke, *On the Expansion of a Coulomb Potential In Spherical Harmonics*, Proc. Cambridge Phil. Soc., 46 (1950), pp. 626-633.

[11] J. Carrier, L. Greengard, and V. Rokhlin, *A Fast Adaptive Multipole Algorithm for Particle Simulations*, Technical Report 496, Yale Computer Science Department, 1986 .

[12] A. J. Chorin, *Numerical Study of Slightly Viscous Flow*, J. Fluid. Mech., 57 (1973), pp. 785-796.

[13] J. M. Dawson, *Particle Simulation of Plasmas*, Rev. Mod. Phys., 55 (1983), pp. 403-447 .

[14] M. Danos and L. C. Maximon, *Multipole Matrix Elements of the Translation Operator*, J. Math. Phys., 6 (1965), pp. 766-778 .

[15] P. G. Debrunner and H. Frauenfelder, *Dynamics of Proteins*, Ann. Rev. Phys. Chem., 33 (1982), pp. 283-299 .

[16] A. Gerasoulis, M. Grigoriaddis and Liping Sun, *A Fast Algorithm for Trummer's Problem*, LCSR-TR-77, Department of Computer Science, Rutgers University, New Brunswick, (1985).

[17] A. Gerasoulis, *A Fast Algorithm for the Multiplication of Generalized Hilbert Matrices with Vectors*, LCSR-TR-79, Department of Computer Science, Rutgers University, New Brunswick, (1986).

[18] G. Golub, *Trummer Problem*, SIGACT news 17(1985), No.2, ACM Special Interest Group on Automata and Computability Theory, p. 17.2-12.

[19] I. S. Gradshteyn and I. M. Ryzhik, *Tables of Integrals, Series, and Products*, Academic Press, New York, 1980 .

[20] L. Greengard and V. Rokhlin, *A Fast Algorithm for Particle Simulations*, Technical Report 459, Yale Computer Science Department, 1986 .

[21] L. Greengard and V. Rokhlin, *A Fast Algorithm for Particle Simulations*, J. Comput. Phys., to appear .

[22] P. Henrici, *Applied and Computational Complex Analysis, vol III*, Wiley, New York, 1986 .

[23] E. W. Hobson, *The Theory of Spherical and Ellipsoidal Harmonics*, Chelsea, New York, 1955 .

[24] R. W. Hockney and J. W. Eastwood, *Computer Simulation Using Particles*, McGraw-Hill, New York, 1981 .

[25] O. D. Kellogg, *Foundations of Potential Theory*, Dover, New York, 1953 .

[26] M. Levitt, *Protein Conformation, Dynamics, and Folding by Computer Simulation*, Ann. Rev. Biophys. Bioeng., 11 (1982), pp. 251-271 .

[27] J. C. Maxwell, *Treatise on Electricity and Magnetism, 3rd edition (1891)*, Dover, New York, 1954 .

[28] J. A. McCammon and M. Karplus, *Simulation of Protein Dynamics*, Ann. Rev. Phys. Chem., 31 (1980), pp. 29-45 .

[29] I. R. Mcdonald and K. Singer, *Calculation of Thermodynamic Properties of Liquid Argon from Lennard-Jones Parameters by a Monte Carlo Method*, Discuss: Faraday Soc., 43 (1967), pp. 40-49 .

[30] M. Neumann, *The Dielectric Constant of Water. Computer Simulations with the MCY Potential*, J. Chem. Phys., 82 (1985) pp. 5663-5672 .

[31] P. van der Ploeg and H. J. C. Berendsen, *Molecular Dynamics Simulation of a Bilayer Membrane*, J. Chem. Phys., 76 (1982) pp. 3271-3276 .

[32] G. Polya and G. Latta, *Complex Variables*, Wiley, New York, 1974 .

[33] A. Rahman, *Correlation in the Motion of Atoms in Liquid Argon*, Phys. Rev., ser. A, 136 (1964), pp. 405-411 .

[34] A. Rahman and F. H. Stillinger, *Molecular Dynamics Study of Liquid Water*, J. Chem. Phys., 55 (1971), pp. 3336-3359 .

[35] L. Reichel, *A Matrix Problem With Applications to the Rapid Solution of Integral Equations*, Report, Department of Mathematics, University of Kentucky, Lexington, 1986.

[36] V. Rokhlin, *Rapid Solution of Integral Equations of Classical Potential Theory*, J. Comput. Phys., 60 (1985), pp. 187-207.

[37] V. Rokhlin, *Application of Volume Integrals to the Solution of Partial Differential Equations*, Comp. and Maths. with Appls., 11 (1985), pp. 667-679.

[38] M. E. Rose, *The Electrostatic Interaction of Two Arbitrary Charge Distributions*, J. Math. and Phys., 37 (1958), pp. 215-222 .

[39] M. J. L. Sangster and M. Dixon, *Interionic Potentials in Alkali Halides and their Use in Simulations of the Molten Salts*, Adv. in Phys., 25 (1976), pp. 247-342.

[40] M. Trummer, *An Efficient Implementation of a Conformal Mapping Method Using the Szegö kernel*, SIAM J. Numerical Analysis, 23(1986), pp. 853-872.

[41] P. R. Wallace, *Mathematical Analysis of Physical Problems*, Dover, New York, 1984 .

Index

The MIT Press, with Peter Denning, general consulting editor, and Brian Randell, European consulting editor, publishes computer science books in the following series:

ACM Doctoral Dissertation Award and Distinguished Dissertation Series

Artificial Intelligence, Patrick Winston and Michael Brady, editors

Charles Babbage Institute Reprint Series for the History of Computing, Martin Campbell-Kelly, editor

Computer Systems, Herb Schwetman, editor

Exploring with Logo, E. Paul Goldenberg, editor

Foundations of Computing, Michael Garey and Albert Meyer, editors

History of Computing, I. Bernard Cohen and William Aspray, editors

Information Systems, Michael Lesk, editor

Logic Programming, Ehud Shapiro, editor; Fernando Pereira, Koichi Furukawa, and D. H. D. Warren, associate editors

The MIT Electrical Engineering and Computer Science Series

Scientific Computation, Dennis Gannon, editor

www.ingramcontent.com/pod-product-compliance
Lightning Source LLC
Chambersburg PA
CBHW060450060326
40689CB00020B/4485